MW01052421

THE WAKE OF CROWS

CRITICAL PERSPECTIVES ON ANIMALS

CRITICAL PERSPECTIVES ON ANIMALS: THEORY, CULTURE, SCIENCE, AND LAW

Series Editors: Gary L. Francione and Gary Steiner

The emerging interdisciplinary field of animal studies seeks to shed light on the nature of animal experience and the moral status of animals in ways that overcome the limitations of traditional approaches. Recent work on animals has been characterized by an increasing recognition of the importance of crossing disciplinary boundaries and exploring the affinities as well as the differences among the approaches of fields such as philosophy, law, sociology, political theory, ethology, and literary studies to questions pertaining to animals. This recognition has brought with it an openness to rethinking the very terms of critical inquiry and the traditional assumptions about human being and its relationship to the animal world. The books published in this series seek to contribute to contemporary reflections on the basic terms and methods of critical inquiry by focusing on fundamental questions arising out of the relationships and confrontations between humans and nonhuman animals, and ultimately to enrich our appreciation of the nature and ethical significance of nonhuman animals by providing a forum for the interdisciplinary exploration of questions and problems that have traditionally been confined within narrowly circumscribed disciplinary boundaries.

The Animal Rights Debate: Abolition or Regulation?, Gary L. Francione and Robert Garner

Animal Rights Without Liberation: Applied Ethics and Human Obligations, Alasdair Cochrane

Experiencing Animal Minds: An Anthology of Animal-Human Encounters, edited by Julie A. Smith and Robert W. Mitchell

Animalia Americana: Animal Representations and Biopolitical Subjectivity, Colleen Glenney Boggs

Animal Oppression and Human Violence: Domesecration, Capitalism, and Global Conflict, David A. Nibert

Animals and the Limits of Postmodernism, Gary Steiner

Being Animal: Beasts and Boundaries in Nature Ethics, Anna L. Peterson

Flight Ways: Life and Loss at the Edge of Extinction, Thom van Dooren

Eat This Book: A Carnivore's Manifesto, Dominique Lestel

Beating Hearts: Abortion and Animal Rights, Sherry F. Colb and Michael C. Dorf

THE WAKE OF CROWS

LIVING AND DYING IN SHARED WORLDS

Thom van Dooren

Columbia University Press
New York

Columbia University Press
Publishers Since 1893
New York Chichester, West Sussex
cup.columbia.edu

Library of Congress Cataloging-in-Publication Data
Names: Van Dooren, Thom, 1980- author.
Title: The wake of crows : living and dying in shared worlds / Thom van Dooren.
Description: New York : Columbia University Press, [2019] | Series: Critical perspectives on animals: theory, culture, science, and law | Includes bibliographical references and index.
Identifiers: LCCN 2019006300 (print) | LCCN 2019008495 (ebook) |
ISBN 9780231544399 (e-book) | ISBN 9780231182829 (cloth)
Subjects: LCSH: Crows—Ecology. | Human-animal relationships.
Classification: LCC QL696.P2367 (ebook) | LCC QL696.P2367 V36 2019 (print) |
DDC 598.8/64—dc23
LC record available at https://lccn.loc.gov/2019006300

Cover design: Milenda Nan Ok Lee
Cover image: Suzanne Dehne / © Getty Images

Earlier versions of chapters 2 and 3 were previously published as:

Thom van Dooren, "Spectral Crows in Hawai'i: Conservation and the Work of Inheritance," in *Extinction Studies: Stories of Time, Death and Generations*, eds. Deborah Bird Rose, Thom van Dooren, and Matthew Chrulew (New York: Columbia University Press, 2017).

Thom van Dooren, "The Unwelcome Crows: Hospitality in the Anthropocene," *Angelaki: Journal of the Theoretical Humanities* 21 (2016): 193–212, www.tandfonline.com/doi/abs/10.1080/0969725X.2016.1182737.

For Deb, an inspiring teacher and a cherished friend

CONTENTS

ACKNOWLEDGMENTS

This book has been the primary focus of my thinking and writing for roughly the past six years. During this time I have undertaken fieldwork in various parts of the world, had two primary academic homes, and been fortunate enough to be a visiting fellow and to present this work at a range of universities. Many people have contributed significantly to this project and to the book that it has become.

I am sincerely thankful to the colleagues who made the time to read and provide feedback on draft chapters. This commentary has been invaluable and has shaped my thinking profoundly. Particular thanks to Michelle Bastian, Etienne Benson, Brad Bolman, Veit Braun, Brett Buchanan, Matthew Chrulew, Margaret Cook, Eileen Crist, Thibault De Meyer, Vinciane Despret, David Farrier, Franklin Ginn, Margret Grebowicz, Donna Haraway, Matthew Kearnes, Lindsay Kelley, Eben Kirksey, Jean Langford, Jamie Lorimer, Stephen Muecke, Ursula Münster, Astrida Neimanis, Emily O'Gorman, Craig Santos Perez, Hugo Reinert, Harriet Ritvo, Deborah Bird Rose, Isabelle Stengers, and Anna Tsing.

I would also like to thank the many talented PhD and honors students that I have worked with while completing this book. Our discussions have often taken me in productive new directions, and I have benefited immensely from all of your inspiring work: Hélène Ahlberger Le Deunff,

Emily Crawford, Kate Judith, Laura McLauchlan, Jamie Wang, and Sam Widin.

This book is, in very large part, a product of my time as part of the Environmental Humanities group at the University of New South Wales. My many wonderful colleagues in this group challenged and provoked me; they introduced me to a range of new ideas and deepened my understandings of old ones. Toward the end of my time on this project I moved to the University of Sydney, where my new colleagues, in particular those in the Department of Gender and Cultural Studies and the Sydney Environment Institute, provided a stimulating and supportive home for the final stages of the work on this book.

During this period of research (2013–2018) I have also been fortunate enough to be able to spend extended periods of time as a visitor at the Rachel Carson Center for Environment and Society in Munich (2014–2016, intermittent), the KTH Environmental Humanities Laboratory in Stockholm (2014), the Anthropology Department at MIT (2018), and the Center for Pacific Islands Studies at the University of Hawaiʻi at Mānoa (2018). Thank you to my hosts—Christof Mauch, Sverker Sörlin, Stefan Helmreich and Heather Paxson, and Alex Mawyer—and their many colleagues for conversations, reading groups, and discussions that enriched this work in so many ways.

This book has also benefited from conversations and presentations of draft chapters during shorter visits—from a couple of days to a couple of weeks—at many universities and other institutions around the world, often alongside or on my way to fieldwork. In particular, I would like to thank my hosts at Aarhus University, the American Museum of Natural History, Columbia University, Curtin University, École Normale Supérieure Paris, UCLA, UC Santa Cruz, the University of Edinburgh, the University of Leeds, Université Libre de Bruxelles, the University of Oxford, and the University of Pennsylvania.

The fieldwork that informs this book has taken me to numerous parts of the globe and brought me into conversation with many fascinating people. I can't hope to name them all here but would like to offer particular thanks to some of those who went above and beyond to show me around, to tell me their stories, or to help me in countless other ways: in Brisbane: Darryl Jones, Trixie Benbrook, Kristen Dangerfield, and Matt Brown; in Hawaiʻi: Hannah Kihalani Springer, Mike Hadfield, Sheila Conant, Donna Ball,

Paul Banko, Sam ʻOhu Gon III, Rich Switzer, Alan Lieberman, Nohea Kaʻawa, John Replogle, and Shalan Crysdale; in Hoek van Holland: Sabine Rietkerk, Harm Niesen, and Colin Ryall; in the Mojave: Tim Shields, Bill Boarman, and Roy Averill-Murray; on Rota: Stan Taisacan, Thomas Mendiola, Sarah Faegre, Phil Hannon, Beth Chagnon, Dacia Wiitala, Doug Page, and Earl Campbell. Finally, Nicola Clayton and Thomas Bugnyar both made time in their busy schedules, more than once, to meet with me and talk about corvid behavioral research. Thank you to you all.

The fieldwork and visiting positions that inform this book would not have been possible without the generous funding provided by the Australian Research Council (DP150103232) and the Alexander von Humboldt Foundation, Germany.

The wonderful staff at Columbia University Press has, as always, made the publication process a seamless one. My particular thanks to Wendy Lochner for her guidance and insight and to Lowell Frye, Michael Haskell, and Robert Fellman for their support and careful attention to detail.

Finally, I would like to acknowledge Emily O'Gorman, my wife, for her ongoing support and for the many inspiring conversations that guided me on my way.

THE WAKE OF CROWS

Introduction

MAKING WORLDS WITH CROWS

I don't know when it was that I first became aware that crows were watching me. Or rather, that as I watched them, they watched back. I grew up and have lived most of my life in places dominated by a particular crow species: the Australian raven (*Corvus coronoides*). Here, as throughout this book, I use the term *crow* to refer to the genus *Corvus*—the so-called true crows—a group that includes the birds generally called ravens, jackdaws, rooks, and crows. While the distributions of the five species of crows in Australia overlap significantly—such that when you look at a bird map most parts of the country should be occupied by two or three species—the reality on the ground (and in the air) is that every major Australian city has come to be dominated by a single species (Rowley 1973). It isn't the same species in every place, but somehow each city has become an almost exclusive territory of one or another species. In the major cities of my life, Canberra and Sydney, that species is the largest of the five, the Australian raven. So I assume that it was these birds who first taught me about crows, that it was under their watchful gaze that I first experienced the sometimes unsettling but always intriguing feeling of being watched by a strange intelligence.

I take this experience of being watched as an invitation to pay attention to a "wakeful world" (Restall Orr 2012), to a world of diverse forms of mindful and creative presence, of beings with their own understandings and desires, their own modes of inquisitive and agentive life. In short, an invitation into what Val Plumwood (2009) called "nature in the active voice." Of course, if we pay attention, all living beings issue an invitation of sorts, as do many nonliving entities and processes.[1] But it has always seemed to me that crows do so in a particularly blatant, perhaps even dramatic manner. For example, if we delve a little deeper into their watchful behavior, things get even more interesting. Watching is not a passive activity, at least as crows do it. Recent research has shown that beyond simply noticing us, they are paying careful attention and acting on this information: they are calculating, sizing us up, adjusting their behaviors, even remembering our faces and passing information about us as individuals on to one another (Marzluff et al. 2010). They are also following our gazes: paying attention to what it is that we're looking at, when it doesn't happen to be them (Schloegl, Kotrschal, and Bugnyar 2007; von Bayern and Emery 2009).[2] In short, watching is here indicative of a whole world of creative, attentive responsiveness. In the presence of a crow it is incredibly difficult to pretend to inhabit a world in which all else is passive background to human lives and dramas. If we pay them even the smallest bit of attention, crows burst the anthropocentric bubble with spectacular flair.

THE CROWS' WAKE

This book is an exploration of possibilities for living and dying well with others—human and not—in an increasingly uncertain world. The challenges of our time take diverse forms: from climate change to colonization, from mass extinction to mass consumption, from toxins accumulating in ecosystems and bodies to an ever widening disparity between the rich and the poor. Where once these issues tended to be divided up into the environmental and the social (and, indeed, they still are by many), perhaps one of the hallmarks of our time is a growing awareness that this is no longer possible (something that many people have always known). We live on a "damaged planet" (Tsing et al. 2017). So pervasive is the impact of particular forms of human life over earthly possibilities that our contemporary period is increasingly coming to be called the Anthropocene, the age of humanity (discussed further in chapter 3). But so unequal are these impacts and culpability for them, and so deep their roots in particular forms of social and economic life, that others have quite reasonably insisted that "the Capitalocene" or "the Plantationocene" might offer a more apt nomenclature for our current predicament (Moore 2015, Haraway 2015).

This book is a very particular take on the messy challenges of our contemporary period, one in which crows guide the way. It takes the form of a series of five stories, each one situated in a different part of the world, in a particular set of human-crow relationships. These stories weave my own travels, interviews, and observations together with the diverse literatures of the humanities and biology to explore how specific lives and deaths are taking form in these places and in each case to ask what else might be possible. At its core, this book is an effort to imagine and put into practice a *multispecies ethics*, one that takes up the monumental challenges of our time through a detailed focus on some very particular crows and their people. In this way, it asks what ethics might look like if we take crows seriously: not just what they *need* but what they can *do*. That is, not just as possible *subjects* of others' ethical regard but as beings who are themselves shaping our shared worlds in consequential ways.

Crows are, in many ways, ideal guides. As a virtually ubiquitous presence around the world—from the arctic to desert landscapes, from tiny islands to the largest continents, and throughout urban, rural, and "wilderness" areas—crows offer a remarkable diversity of instructive sites for thinking. Indeed, crows of one sort or another are to be found across most of the world's continents (the exceptions being South America and Antarctica), and on many smaller islands (Haring et al. 2012). But they are also tangled up with people in diverse ways. The crows with which most of us are familiar are those species that thrive on human waste and have taken up residence in cities and on farms. Few other animals have done as well in human-dominated places as crows. But to thrive in these "emergent ecologies" (Kirksey 2015), to do well in this way, is often to suffer from a host of negative associations: around the world crows are often blamed, rightly or wrongly, and persecuted for a variety of impacts on other species and on people's lives and livelihoods. Alongside these more familiar stories, there is also a far less visible cohort of crows—predominantly island species, forest and fruit specialists—that have been pushed to the edge of extinction by various human activities (and in some cases over that edge).

As such, around the world and throughout this book, we encounter crows engaged in a range of relationships with people: some are deeply loved and passionately conserved—perhaps they are vital seed dispersers for endangered forests—while others are predominantly viewed as pests, scavengers on the detritus of human life, and maybe even agents of extinction through their actual or potential impacts on other species. All, in their own ways, however, are threatened—as individuals and sometimes as species, too. Indeed, the crow stories presented in this book fall broadly into two groups. On the one hand are the critically endangered crows and on the other are the crows viewed, by some at least, as overabundant or out of place and so subject to diverse practices of control, killing, and perhaps even extermination.[3]

While the stories in this book all start with and keep coming back to the crows, they are, at the same time, explorations of globalization and the expanding footprint of global trade, of a changing climate, of colonization and associated struggles for Indigenous sovereignty and local autonomy, of entrenched patterns of unequal urbanization and economic development, and of militarisms and their diverse effects on peoples and places. They are stories about shared lives that cut across imagined divisions between "environmental" and "social" domains. For some readers of this book there will likely be too much about crows. For others there will likely be too much about people. To these concerns I can only reply that this book is grounded in the conviction that the challenges of our current period demand modes of thought and attention that work across these kinds of divisions to acknowledge rich and consequential patterns of entanglement. In taking up these topics, this is a book that explores the way in which vital questions of justice, of unequal positionalities and exposures, are inescapably caught up in and woven through with multispecies relationships. Certainly crows, but also many people, live and die, flourish or wither, inside these relationships.

Through a detailed focus on five specific sites and minor forays into several others, this book explores a small part of the diversity of contemporary human-crow entanglements. Chapter 1 focuses on the Torresian crow (*Corvus orru*) in the Australian city of Brisbane, a sometimes loved and sometimes vilified urban presence. The practices of urban ecologists and a particularly dedicated wildlife carer guide the possibilities for cohabitation that emerge here. Chapter 2 explores the complex and entangled forms that absence takes through the figure of the critically endangered Hawaiian crow (*C. hawaiiensis*) in the forests of the island of Hawai'i.

Here contestations between conservationists and pig hunters, which draw in Kānaka Maoli (Native Hawaiians) in diverse ways, are the central focus. Chapter 3 takes us to the small coastal town of Hoek van Holland in the Netherlands, where the local government is diligently working to eradicate a small population of house crows (*C. splendens*) who likely arrived in the area as stowaways on one of the many cargo ships arriving into the Port of Rotterdam. This is a story about the ways in which global trade, industrialism, and mass consumption touch down in particular places. Chapter 4 tells the story of some high-tech projects that aim to prevent the common raven (*C. corax*) from eating threatened tortoises in the Mojave Desert in the United States. Decades of urban sprawl have transformed this place for everyone, drastically changing the possibilities for ongoing life. Finally, chapter 5 takes us to the island of Rota to explore contestations over the future of another endangered Pacific Island corvid, the Mariana crow (*C. kubaryi*). This is a story of shifting relationships among crows, trees, Chamorro people, and outside conservationists, in which the rights of local people to make lives for themselves in their ancestral lands are at stake.

Across these sites we encounter crows living very different kinds of lives. They eat different things, they inhabit different environments, they have very different conservation statuses, and they will likely have sharply divergent futures. But in other ways they are also remarkably similar. The birds referred to as crows in this book are the members of the genus *Corvus* (crows, ravens, rooks, and jackdaws). These birds are all closely related; in most cases the particular common name that each of them has is more an accident of history than a reflection of any meaningful taxonomic difference. For example, those species that are called "ravens" are usually the bigger members of the genus in any given area and are not necessarily any more closely related to one another than they are to other members of the genus. The crows are predominantly black in color, with about a quarter of the world's species also having some grey or white. They are omnivorous birds, although species (and individuals) differ in the variability of their diets.[4]

Around the world, these birds also have in common the fact that they share many of the same complex cognitive, emotional, and social capacities, although, as we will see, they are actually more like "potentialities" than static capacities, each emerging within a particular relational milieu (Despret 2008b).[5] Crows are, to put it simply, all very clever birds, at least by the standards that contemporary science uses to measure these things. In fact, they are now generally thought to be one of the most intelligent groups of animals on the planet (Emery 2017). This understanding is based largely on cognitive and behavioral research that has really only taken firm shape in the last few decades, exploring areas ranging from tool use and tool making to "theory of mind," insight and mental scenario building, future-oriented thinking, and the possession of a sense of fairness. These remarkable capacities are a key part of the stories I tell in this book. They help us appreciate the crafty and creative ways in which crows make sense of and inhabit their worlds. These stories aim to unsettle the all too frequent, simplistic, singular association of crows with death and scavenging: crows, as we will see, are many things, all at the same time. But exploring corvid capacities in this way is also a vital part of understanding why these birds are sometimes such difficult neighbors and of understanding how it is that diverse peoples might still be able to craft more livable possibilities with them.[6]

Ultimately, this book is an effort to think through the complexity of our world-remaking epoch. But it is not a discussion of planetary problems and planetary solutions in the abstract. Nor is it a call to localize, to focus on smaller-scale problems and responses. It is rather an effort to subvert scale, to pay careful attention to some very particular places, their people, and their crows as a way of grounding, making sense, and responding responsibly to processes and problems that are often staggering in their immensity. In short it offers a set of particular stories for particular places, and crows, as we will see, are very particular kinds of creatures. In this way, the book aims to grapple with substantial questions while refusing abstraction, to cultivate ideas, approaches, and modes of attention *with others* in a way that might enable us to see and be differently (see chapter 1). It works toward tentative, collaborative, humble ways of understanding that are only as big as they absolutely need to be, rather than grand interventions. As Donna Haraway (2016, 101) notes: "We need stories (and theories) that are just big enough to gather up the complexities and keep the edges open and greedy for surprising new and old connections."

Through these stories this book works to draw its readers into the crows' wake. The "wake" I have in mind here is intended to capture three interrelated processes and possibilities. In one way or another, all of the stories told in this book center on the killing of crows—extinction, extirpation, extermination. In this context, a wake is a gathering, a pause to reflect in the presence of death. It is an opportunity to grieve and to learn, something that crows themselves seem to be doing in the face of death.[7] Importantly, a "wake of crows" is not a "murder of crows." While both terms carry with them deathly associations, the former also holds within it all manner of possibilities for creative, reparative, collective world making. Indeed, a wake is also an opportunity to celebrate a life and to move forward *well* with those who remain. In approaching death and dying in this way—as a space of simultaneous mourning and celebration—this book, like much of my previous work, seeks to create openings into wonder and appreciation for diverse, even if threatened and disappearing, lifeforms and forms of life. The wake is a space, a possibility, outside of binary oppositions between optimism and despair (see chapter 5). Second, a wake is also a disturbance, a trail left by the movement of something, perhaps a crow, through the air (or water). In this sense it asks us to consider the aftermath, the consequences, of all this killing. But a wake does not just happen *afterward*; it is a dynamic unfolding, emerging as movement takes place. As such, it asks for a relational, ecological engagement with what particular forms of life and death mean and why they matter: it asks us to consider what it means to be swept up with and transformed by one another. Each wake is utterly unique, the result of the very particular interplay between bodies and their atmospheres. Each takes its own individual but thoroughly relational form. Finally, the term evokes the possibility of wakefulness. The ravens with which I open this book invite us into—they awaken an awareness and appreciation of—a wakeful world, a world composed of diverse forms of mindful, agentic, purposeful existence. In this way they invite us into the need for a different kind of ethics, one that might provide new possibilities for understanding and inhabiting genuinely shared worlds, worlds crafted with and for diverse forms of being. Taken together, these three wakes capture the key foci and approach of this book: a set of stories told in the wake of crows.

MULTISPECIES ETHICS

A multispecies ethics is one that takes seriously the fact that all life, including human life, occurs within fundamental and constitutive relationships with other kinds of beings, living and not. Ours are richly diverse multispecies worlds. My use of the term "worlds" in the plural—here and in the title of this book—is grounded in an acknowledgment that we inhabit "a world of many worlds" (Blaser and de la Cadena 2018). Such an understanding works outside dominant Western notions that there is a single "reality" over the top of which diverse "perspectives," "cultures," and "ideas" are simply layered, more or less accurately capturing what lies beneath, unchanged. There are many other ways of understanding, inhabiting, and ultimately "doing" worlds. Our worlds are not preexisting, static entities. They are becomings that must be put together—from the inside—by, through, as the embodied imaginings, presences, and intra-actions of innumerable beings and forces.[8] These are processes of "worlding," of worlds in the making. Of course, to say that worlds are made is not to imply that any one of us—or any coalition that we might form—can unilaterally "decide" how they will be made. But it is to acknowledge the various forms of agency, of very real even if always thoroughly constrained and unequal influence that each of us—and not just the human "us"—has in the shaping of what is. In this regard, ideas, understandings, and modes of meaning making don't float above reality; they *matter*: they are woven through worlds in constitutive ways (Haraway 1997).

To insist that we make and inhabit *worlds* (in the plural) is not to deny a common space of existence. Rather, it is to hold such an idea in tension with another, namely, with the sense that each of us, individually and collectively, also crafts and inhabits distinctive spaces of existence. The ravens that regularly visit the tree outside my window do not experience, understand, and make sense as I do. We inhabit different worlds, populated by different entities, in different relationships. To call these differences "worlds" is to insist that they do not simply reside "in the head" of a human, a raven, or another but rather emerge in relationship with others and ripple out to take material form—more or less consequentially (for whom?)—in the lives these beings live and the ways in which they work to shape the present and the future. Holding these two ideas in tension—that our worlds are common yet distinctive, only partially overlapping—we might say that they exist in the "multiple" (Mol 2002): "More than one and less than many" (Strathern 1991, 35). Throughout this book I have used the notion of a "shared world" or "shared worlds" to describe this reality. To "share" is both to hold in common and to divide something up. In addition to the unavoidable reality of this tension in the ongoing crafting of worlds, I also use this notion to name the ongoing work of crafting worlds that are attuned to and accommodating of their simultaneous distinctiveness and commonality.

There is a growing body of work exploring the particular form that ethics takes when it must be practiced from within worlds in the making, worlds in the multiple.[9] Here and throughout this book I draw on discussions taking place at the intersections among feminist science studies, multispecies studies, anthropology, Indigenous studies, and the environmental humanities.[10] In this work, as Cecilia Åsberg (2013, 2) notes, "processes of 'becoming with' are always also . . . relations of obligation." Here, ethics becomes the work of crafting flourishing worlds, of *worlding well*. Of course, this is not an innocent definition: what counts as "well," as "flourishing," or, indeed, as "crafting" and who is relevant and involved in determining these things are precisely what is at issue. There can be no generalizable or final answers to these questions. Rather, ethics is about what Haraway (2016) has called "staying with the trouble." This is a process of exploring and inhabiting complexity, insisting that there is no singular ideal state but rather only shifting, specific, and always relational possibilities for better and worse forms of being. As such, it is an effort to multiply perspectives on what will count as a better world (and for whom), while owning up to the fact that there is never a situation in which everyone wins—yet actions must still be taken (see chapter 1). As John Law (2015, 128) argues, in such a context, we will need to craft approaches that are "contingent, modest, practical, and thoroughly down-to-earth." In a shared world we cannot assume that any way of understanding, ordering, or valuing is correct or final. Rather, we are required to *attend* to others in their specificity, to ask and re-ask *with* them: What matters here and what else might be possible?[11]

This book takes up and explores this work of *attentiveness*—of paying attention and attending to the complex realities of actual corvids in consequential relationships with humans and a diverse range of others. As Deborah Bird Rose (2007, 88) notes, "Located in the real here and now of encounter," the practice of paying attention "takes us beyond our previously known worlds." It is a way of "putting knowledge at risk and of allowing others, of all shapes and sizes, to make a difference to the process of knowing," and so of being (Hinchliffe et al. 2005, 653). In the small Dutch town of Hoek van Holland, local authorities have invested considerable time and energy in their efforts to eradicate a tiny population of introduced house crows, worried that they might one day multiply and spread out into Europe (chapter 3). Meanwhile, on the island of Hawai'i, a decades-long program of captive breeding seems finally to be paying off with the reintroduction of about twenty critically endangered Hawaiian crows back into the forest (chapter 2). These two crow species are here understood and responded to in sharply divergent ways. But this contrast is more complex than a simple narrative of one crow cherished and another despised. There is no single response to either of these two crow (re)introductions. In Hoek van Holland, some local activists have opposed the killing of these crows through the courts and public protest, even threatening to safeguard some of the birds in their homes to release them at a later date, crows they have referred to as *onderduikers*, with deliberate allusions to the Dutch protection of Jews and other persecuted peoples during the Second World War. Meanwhile, other locals have supported this eradication plan, justifying it, at least in part, through reference to the many other forms of avian diversity that might someday be negatively affected by these crows. In Hawai'i, the presence of crows is an even more fractious issue, highlighting divisions among and within groups of conservationists, hunters, and Kānaka Maoli over land management, which is always powerfully connected to the ongoing realities of the U.S. colonization of these islands.

Worlds are at stake in the ways in which crows and people understand and respond to each other in these places. But they are at stake in very different ways. Different histories, different imagined futures, different ecologies, and different configurations of life and death animate these sites and their possibilities. In this context this book aims to take up the slow, careful work of attending to the particular, of asking what it means to craft flourishing worlds *here*, in this place and time. This is an approach that is prompted and guided *from the outset*—both in the questions asked and the responses offered—by situated human-crow encounters: How do these specific entanglements give rise to possibilities for life and death, and for whom? How do diverse ways of knowing and valuing (human and not) create particular relationships and accountabilities? Here, ethics emerges out of detailed engagements with the specificity of actual worlds in the making. This is an *emergent* ethics, not an *applied* one. As such, this book does not aim to produce an ethical "system" that can be easily put to work in disparate contexts, an "off-the-shelf ethics" (Ginn et al. 2014, 113). Instead, the ethical theorizing undertaken here is inseparable from the particular: theorizing as (re)description, a theory that takes the form of an effort to "redescribe something so that it becomes thicker than it first seems" (Haraway and Goodeve 2000, 108), an effort to weave the world together differently, in new ways, to produce alternative understandings and so possibilities.[12] It is an approach that recognizes that "stories are the real homes of so-called thick moral concepts, concepts in which evaluation and description are so intertwined as to be conceptually inseparable" (Cheney and Weston 1999, 130).

Doing ethics in this way is an ongoing, collaborative, and always speculative meditation on flourishing, on what it means to imagine and craft better worlds with others. This terminology of "flourishing" draws on Haraway's (2008) engagement with the work of the feminist philosopher Chris Cuomo (1998). This term does important multiplication work for these theorists: it moves us away from ethics as an abstract inquiry into "the good" and toward an insistence that there are only ever better and worse worlds and that even these determinations must always be relational and situated. I understand one of the key values of this term to reside in precisely this openness: flourishing offers no fixed currency of ethical consideration, no fixed mode of obligation, no fixed class of ethically relevant beings (Kearnes and van Dooren 2017). Instead, grappling with the multiplicity of worlds in the making requires taking seriously diverse modes of being, agency, and understanding. For some of the people whose stories I engage in this book, flourishing takes the form of corvid conservation, of a world in which threatened crow species are restored. For others, it must involve corvid eradication or at the very least large-scale destruction: the removal of a pest or a threat. A broad range of other "goods" are vested within each of these priorities: economic development and prosperity, Indigenous sovereignty, expanded global trade, the conservation of yet other species affected in one way or another by crows. To ask after flourishing is to be drawn into these complex and often conflicting webs of understanding and relating.

But a multispecies ethics must go much further still. In this case, it must also ask what it means for particular corvids to flourish: what constitutes a good life for a crow in the Mojave Desert or Hawai'i? What kinds of relationships—with other crows, with a wider environment, with local people—might enable or disable these possibilities? Beyond humans and crows, there are also numerous other living beings whose possibilities are caught up and at stake in these worlding processes.[13] In the Mojave Desert, some conservationists are pushing for the ongoing culling of large numbers of common ravens to protect threatened juvenile desert tortoises (*Gopherus agassizii*), whose soft shells are vulnerable to sharp raven beaks. And in Hawai'i, the loss of this endangered crow might have profound consequences for a group of plant species that have become dependent on these birds for the dissemination of their seeds. Exploring questions of worlding well within these inescapably multispecies contexts also requires learning to see and to see-with these others.

A multispecies ethics aims to take seriously a diversity of competing understandings, possibilities, and priorities; a diversity of worlds. Doing so, however, ultimately requires the acknowledgment that flourishing always "involves a constitutive violence . . . some collectives prosper at the expense of others" (Ginn et al. 2014, 115). Despite these challenges, an account must be offered, decisions made. Common ravens in the Mojave will either be culled or they won't; Hawaiian crows and the plants that rely on them either will be supported by conservation efforts or they will perish. The complexity of real worlds and the unavoidable partiality of our understanding cannot be an excuse for indifference and inaction (Shotwell 2016). Multiplying perspectives, the approach offered in this book refuses to allow any "good" simply to trump the others, perhaps on the basis of a generalized claim that human welfare, individual animal suffering (Singer 1975), or endangered species conservation (Rolston 1999) ought to be given the highest priority.[14] Taking all ethical claims seriously means adopting something like a stance of "universal consideration" (Birch 1993) from which to ask what these interactions mean *here*, in this place, for *all* of those involved.[15] On the island of Rota, for example, the conservation of the Mariana crow has delayed the allocation of homestead lands that would have enabled Indigenous Chamorro people to begin farming, to begin to imagine and craft other futures for themselves, giving rise to anger and resentment among locals who as a result target the crows. This is a poor outcome for everyone. Doing ethics, attending to worldings, is not simply about which claims should take priority. It is, rather, about exploring the histories and imagined futures that have given rise to this contestation and that will circumscribe the crafting of alternatives. It is about a careful attention to what these various scenarios might mean for local people, for crows, for the forest, and for many others, as well as to what else might still be possible, how we might all become *differently*, together.

Crows and their humans are my focus: it is their possibilities for life and death that centrally occupy me. A focus of some sort is an unavoidable reality of all analyses, all modes of sense making. This does not change the fact that focusing is high-stakes work: the particular focus I have taken in each of these sites has shaped what I have been able to see. Each of the stories I tell here might have been told in countless other ways, layering in other voices, other possibilities, other worlds, that haunt the edges of my own accounts or are perhaps missing entirely. Where are the thick and where are the thin presences? Which lives are inadequately regarded or might have been thought otherwise with the benefit of greater attention? These are vital and unavoidable questions. Ultimately, however, they cannot be resolved away and instead require ongoing attention. They require the kind of situating work that acknowledges that ways of seeing are never neutral or innocent and as such that the goal is not to render them so but to learn to see better their particular omissions and violences and to become accountable for them where we can. The partiality of any assembled understanding is an invitation to openness, to connecting and learning to see *with* others (Haraway 1991, 590). But there will always be more to learn, more stories and perspectives that might complicate or even redo what we thought we knew. As such, the partiality of perspectives is also an invitation into an ethics grounded in uncertainty, in humility, in open-ended questioning. This does not mean that ethical decisions can never be arrived at—indeed, they *must*—but instead that there will be no final or singular ethical outcome: the good must be carefully crafted, in the multiple, again and again. A restless ethics for a wakeful world.

In the stories that occupy the pages of this book, I take up this obligation as an effort to *find a way through* the complexity of voices and claims, of worlds, however tentative and compromised it might be. *Multiplying perspectives is not an ethics.* It is a vital *part* of a situated ethics of worlding, but it cannot be an end in itself (Giraud et al. 2018, 74).[16] My goal in this book has been to tell stories that grapple with multiplicity—with entangled modes of life and competing forms of flourishing, as well as diverse perspectives, histories, and futures—but that ultimately advocate for particular worldly possibilities. These are my own stories: my sense of what it might mean to do worlds well, or at least better, in these contexts. These stories are crafted in conversation with others—they work to be faithful to the diverse people, crows, and others whom I have become tangled up with in the course of researching and writing this book—but ultimately I must be accountable for what is said here: for the worlds that I learn to see and to share. As such, grappling with multiplicity is not an endeavor to tell others' stories but rather simply to tell my own stories in responsible ways (van Dooren and Rose 2016). This telling of stories is also part of the *work* of ethics as worlding: however modest their effects, the stories that we tell play a role in the imagining and shaping of possibilities.

In a context in which we are all, unavoidably, crafting worlds with others, multispecies ethics represents a commitment to *worlding well*, to contributing in whatever ways we can to the imagining and crafting of flourishing, abundant worlds (Collard, Dempsey, and Sundberg 2015a). In Isabelle Stengers's (2014, 8) terms, it is about "accepting that what we add makes a difference in the world and becoming able to answer for the manner of this difference." To this end, each of the following chapters works to "thicken" our understanding of flourishing in its own ways. Chapters explore forms of ongoing openness and experimentation in collective life, the need to attend to histories/inheritances and imagined futures, the value of humility and uncertainty in eschewing an appropriative stance to others, and the need to think ecologically to appreciate the specificity of our being entangled and at stake in one another. Taken together, these chapters do not aim to offer a definitive set of criteria for multispecies ethics but rather to explore some of the many questions, attentions, and openings that might be practiced or kept in mind in situated efforts to ask what flourishing might mean in specific places and times.

FIELD PHILOSOPHY (WITH KEYWORDS)

Exploring these questions in a serious way is not something that can be done from an armchair—or indeed from any distant location. Instead, it requires modes of research and/as engagement that take us out into "the field." It requires a "field philosophy" (Buchanan, Bastian, and Chrulew 2018). The stories I tell in this book, as with much of my previous work, draw on field research—principally interviews and my own observations—with some of the many people whose lives are caught up with crows in one way or another: conservationists, farmers, hunters, Indigenous peoples, and many others. At the same time, I also sought to better understand the crows, their social relationships, their cognitive and emotional lives, their ecologies: How do they make sense of the world? What matters to them and why? To this end, my travels took me to some of the world's leading laboratories of corvid behavioral biology as well as out into the field with people studying or living with crows in these particular sites. I have drawn on interviews with biologists and the relevant scientific literatures as well as on my own corvid observations and the diverse knowledges of many other people who make it their business to try to better understand crows and their environments.

Each of the chapters in this book draws on this research to tell a story that centers on a particular key concept. In Brisbane, I explore the possibilities for crafting *community* with crows and others; on the island of Hawai'i, I ask about *inheritance* as a work of holding onto and letting go of the past in its various forms; in Hoek van Holland, I consider *hospitality* as a way of welcoming and of sharing placetimes; in the Mojave Desert, I ask about *recognition* and the ways in which both people and ravens make sense of a world of subjects, of other living beings with their own agendas and worth; and, finally, on the island of Rota, I explore possibilities for *hope* and the ways in which crows and people are imagining and working toward particular futures.

Community, inheritance, hospitality, recognition, hope: these are the five keywords that animate this book. Each is a *mode of worlding*, a space of possibility for understanding and enacting worlds, with all of the opportunities and the limitations that this implies. As previously discussed, the productive ambiguity of flourishing enables a kind of ethical inquiry that does not presuppose at the outset who counts and what matters. But in doing so, flourishing also provides little by way of an anchor, a focus around which inquiry might proceed. As such, the approach I adopted in researching and writing this book was to deliberately ground each chapter in a single concept as a way of gathering up and centering my inquiry in each site: what kinds of flourishing are possible within different enactments of community or within different approaches to inheritance or hopefulness?

The specific key concepts explored in this book were not chosen at random. Each was tentatively selected as a locus of interesting and difficult possibilities on the basis of background reading and research and then taken into the field. In most cases they "worked"—although never as expected—but in one case a substitution was necessary; the site suggested a different focus (chapter 4). In each case, thinking through a particular concept helped me see unexpected connections and relationships that would otherwise have been far less obvious; it helped me tell different stories. The challenge, of course, is in avoiding these concepts becoming blinkers while acknowledging that they do inevitably structure what we are able to see and learn. Other concepts would have prompted other stories. As Stephanie LeMenager (2017) has noted, there has been a surge of interest in lexicons and keywords within the environmental humanities and related fields in recent years.[17] To my mind, it is precisely this focusing effect that lies behind this interest: in fields dominated by an impulse to multiply voices, to acknowledge plurality, keywords become concrete foci around which complexity can be rendered in some way intelligible.

These particular keywords might be helpful for multispecies ethics in other sites, in relation to other questions and other species. But they may not be. Nothing can be taken for granted. Much will depend on the specificity of the site: making concepts travel is always a *work*, requiring translation, transformation, and sometimes outright substitution. With this in mind, this book does not aim to provide a lexicon or even a first step toward one; each of these keywords is what Mario Blaser and Marisol de la Cadena (2018) have called a "*concrete* abstraction." More than in any of the particular terms worked with, I see the wider value of the keywords approach taken here to reside in the opening up of a three-way conversation among the environmental humanities, the biological sciences, and the specificity of particular places and cases.

This conversation begins with the distinctive literatures that each of these key terms has in both the humanities and biology. Terms like "community," "inheritance," and "recognition" have quite explicit histories and valences in each of these fields. In contrast, "hospitality" and "hope" have obvious humanities literatures while also inviting conversation with biological literatures on "territoriality" (not to mention the host-parasite relationship) and "prospection" or future-oriented thinking, respectively. But this book is an effort to bring these diverse literatures into creative dialogue *in particular places*, with respect to particular human, corvid, and other lives. In this way, diverse local understandings and valences are also powerful presences in the ways in which these chapters and these concepts take form.

On Rota, I was captivated by Thomas Mendiola's notion of "provisioning" in a way that completely redid my understanding of hope. In Hawai'i, Hannah Kihalani Springer taught me to understand inheritance as a political work through her sense of herself as a "citizen of the land." In Brisbane, Trixie Benbrook introduced me to some of the many ways in which multispecies community might be articulated in practice. To philosophize in the field with others in this way is to explore a diversity of ways of making sense—something that our more ethnographically minded colleagues in anthropology and other disciplines have long known. It is to take seriously the ideas and insights of all of those people—and, indeed, other beings, too—who wonder about and take an interest in understanding our worlds. In so doing, field philosophy has the potential to challenge and unsettle the homogeneity of the Western philosophical canon, at least as it is often taught and thought about, with a predominant focus on the works of educated, white men (Hutchison and Jenkins 2013, Yancy 2007).

In exploring these concepts in these places, I also take seriously the possibility, expressed to me most pointedly by Isabelle Stengers in response to this project, that many of these terms have deeply human-centered histories and connotations. My aim here is not to disavow these connections and meanings but rather to inherit them *otherwise*, to twist and torque these concepts in ways that are productively unfaithful to their (often) humanist origins. Inheritance is, after all, always a question of both retention and transformation (see chapter 2). In so doing, this book asks: What value(s) might these concepts have for multispecies ethics? Can they be repurposed, or are they ultimately unsalvageable? Taking up these questions, I aim to address the serious challenge that, as Claire Colebrook (2012, 185) has put it in her discussion of climate change, "none of the terms of our ethical vocabulary . . . is up to the task. Many terms and principles might even exacerbate the problem."

The biological literatures I engage with in this context are primarily those of behavioral biology and ethology (focused on corvid cognition and behavior), ecology, and conservation biology. As I've noted, I draw on the insights of biology in a general way in this book to develop a "thicker" sense of corvid life—to pay attention to crows—and so to better understand what matters to them and what it might mean to craft flourishing worlds with them. But alongside this engagement, I draw on biology in relation to these specific keywords in an effort to unsettle them productively.

What might research on "theory of mind" in cognitive biology—that is, the capacity that some animals have to recognize other living beings as independent loci of intention and agency—add to an understanding of recognition as an ethical imperative? What might we learn by considering together the various biocultural forms of inheritance—from genes to traditional cultural practices—that constitute bodies and worlds? The point here is not that biology has the answers or that biological understandings can simply be combined with those of the humanities or local peoples, in a straightforward way. Rather, my aim is to pull these understandings into a critical dialogue, to emplace them in a way that might produce new understandings.

One important aspect of doing so is the effort to open up space for the multiplication of creative forms of attention toward other-than-human modes of worlding. How do crows do hope, recognition, hospitality, community, or inheritance? Are these meaningful questions? What might asking them allow us to see and do with others? This book deliberately avoids the more conventional approach to asking about whether (nonhuman) animals might "do ethics," an approach that has centered on fairness studies and other ethological inquiries that count and measure animal behaviors and choices in relation to one another (de Waal 2009). These studies are important; they fascinate me, and I discuss them briefly in this book (see "Cooperating"). In general terms, however, my sense of ethics is more experimental, open ended, and tentative than this. Fundamentally, it is a question of whether and in what ways crows might be taking up the task of more or less intentionally, thoughtfully, generously making worlds with others: of *worlding well*. In each chapter, this question takes a different form. In my discussion of Brisbane's urban crows, I explore this possibility in the context of the work of community. While crows are certainly engaged in crafting relationships and social collectives with various species, are they the kinds of creatures who might do so in a way that deliberately attends to and makes room for these others?—something that I am thinking about as the challenge of a "situated pluralism" (chapter 1). Meanwhile, my discussion of the Mariana Islands centers on the question of whether and how crows might be engaged in hopeful acts: in imagining and deliberately working toward particular futures that they prefer over other possibilities (chapter 5).

My aim in this book is not to settle these questions but rather to open them up as new spaces for attention. To this end, in various parts of this book I consider the possible valences of some of these key concepts beyond the worlds of crows. What other forms might they take in a broader more-than-human world? Val Plumwood (2009) provided the seed for this inquiry in her discussions of "nature in the active voice," which aimed to rethink notions of creativity and mindfulness in an effort to find them in the wider material world rather than view them as the exclusive domain of human or perhaps even "higher-animal" life. For example, in invoking the notion of the "wisdom of the wing," Plumwood refers to a kind of wisdom that we might see residing in the evolutionary process. "Why," she asks, "can't we see evolution . . . as a form of experimentation, of testing and learning, like trial and error, a form of wisdom?" (125). I read Plumwood's proposal here as one of multiplication rather than extension, an invitation to attend to modes of being that contain important similarities (alongside differences, a tension that Plumwood's work was always highly attuned to; see chapters 3 and 4). The point here is not to collapse diversity under the same name and so erase difference—to call everything "wisdom"—but rather to enter into an imaginative engagement with a wider world, to seek out perhaps unfamiliar or even uncomfortable expressions of this possibility, and to ask what learning to approach the world in this way might help us see and appreciate.

STRUCTURE OF THE BOOK

This book comprises five substantive chapters and five short vignettes. Each chapter is grounded in a particular site and key concept. The interspersed vignettes offer some concentrated, albeit brief opportunities to reflect on some of the remarkable features of crow life. Drawing in particular on research in behavioral biology, these vignettes explore what crows might be up to when they experiment with cars as a means of opening tough nuts ("Experimenting"), when they steal from each other ("Stealing"), when they pull a string together to access food ("Cooperating"), when they hold their wings open over a lit cigarette ("Fumigating"), and when they seemingly leave shiny trinkets for friendly people ("Gifting"). In each of these cases, we learn a little more—or at the very least are provided with some fascinating sites for careful speculation—about how corvids make sense of the world. Many of the key ethological concepts and competencies discussed in these vignettes—things like theory of mind, desire-state attribution, caching behavior, and neophobia—are also important elements of the key arguments developed in the book's chapters. In this way, these vignettes also function to enrich and support the more substantive discussions of human-crow intra-actions offered in the chapters.

Chapter 1 takes us to the Australian city of Brisbane to explore some of the ways that Torresian crows are experimenting with new possibilities for life in this changing urban environment. Some of these experiments, it turns out, are not to the liking of at least some local residents, giving rise to complaints and occasional violent reprisals. Focusing in on possibilities for multispecies community, the chapter proposes that these corvid activities might be understood as ethicopolitical interjections that interrupt, propose, and enact forms of shared life. Drawing on archival research and interviews with a range of local people, the chapter explores the way in which ideas about an ideal community—specifically those grounded in the notion of achieving a "natural balance"—rub up against more complex and multifaceted realities. Ultimately, the chapter explores an alternative approach to community as a work of "situated pluralism" that holds permanently open the question of who we are, where we are going, and how we might decide.

Chapter 2 centers on the Hawaiian crow, or ʻalalā, on the island of Hawaiʻi (the Big Island). For roughly fifteen years the ʻalalā was extinct in the wild, the only remaining participants of the species required to live their lives in captivity, subjects of a long-running conservation plan that aimed to breed enough birds for eventual release. Since 2017, birds have at last started being returned to the forest. This chapter draws primarily on fieldwork conducted before these releases to explore contestations over the management of a large section of forest that is a likely future home for ʻalalā. At the center of these contestations was a proposal to regenerate the forest understory through the fencing and exclusion of pigs, animals valued by many people for hunting and seen as central to Kānaka Maoli hunting as a traditional cultural practice in the islands. In this context, this chapter explores how we inherit and inhabit the legacies of the past to shape possible futures. These inheritances take many forms, from genetic material and the broader landscapes and ecological communities that we are born into to the historical events, cultural traditions, and relationships that we retell, reenact, and remember. In a time of ongoing extinction and colonization, a time in many ways characterized by interwoven patterns of biological and cultural loss, what does it mean to inherit *responsibly*?

Chapter 3 focuses on a small population of roughly forty house crows in the town of Hoek van Holland in the Netherlands. As their common name implies, house crows stick pretty closely to people, so much so that there are no known populations living independently of us. The small population of crows at the center of this story are all likely descendants of two or perhaps a few birds that arrived on container ships in the mid-1990s. In 2014, after twenty years of peaceful coexistence, the government of the province of South Holland began the process of eradicating them, worried that they may one day become a pest or threat to biodiversity. Just across the water from Hoek van Holland is the Port of Rotterdam—Europe's largest port—an "engine" for the global patterns of production, trade, and consumption that are today remaking our world, ushering in what many are calling the Anthropocene. Focusing on these crows and this port—in a way attuned to the broader placetimes that constitute our present—this chapter seeks a more situated way into the relatively abstract notion of the Anthropocene. Working through the concept of hospitality, it explores the ways in which other species are made welcome, or not, in the places that we call "our" own. Telling the story of this little group of birds in a way that holds this port and its effects in the frame, this chapter asks how we might be required to rethink our responses to and to learn to live with others in this difficult time.

Chapter 4 takes us to the desert, specifically the Mojave Desert in the southwestern United States, to explore a range of creative projects aimed at preventing the predation of the endangered desert tortoise by the common raven. Working against ongoing calls for large-scale "lethal control," a local group called Hardshell Labs is producing a range of nonlethal technologies, from "weaponized" 3D-printed tortoise shells to drones and lasers and even videogame interfaces to crowdsource labor. Focusing on the question of recognition, this chapter explores the way in which these interventions are grounded in a particular understanding of ravens, that is, an understanding of ravens as *subjects*, as wily, intelligent, adaptive beings. This understanding enables these alternative interventions; it enables possibilities for what Vinciane Despret and Isabelle Stengers have called "diplomacy." But alongside these new possibilities for cohabitation, ravens as subjects also raise the specter of possible new forms of psychic and social trauma. As such, the chapter also asks about the new obligations that might emerge as we come to recognize others as subjects that are also susceptible to distinctive forms of harm.

Chapter 5 centers on the critically endangered Mariana crow, or aga, on the island of Rota. While the causes of the precipitous decline of this species are contested and likely multiple, it seems that direct and deliberate persecution of these birds by Indigenous Chamorro people is a key component of this story. Driven both by frustration and a practical desire to keep crows off their lands—to avoid the conservation restrictions on livelihoods and land practices that are now bound up with these feathery bodies—many local people have taken to killing aga or at the very least removing their nesting and food trees. This chapter explores the interfaces of development and conservation on Rota with a particular eye on hope. The book ends with hope not because I am aiming for an upbeat finale; my understanding of hope is much more ambivalent than this. Rather, as will become clear, I take hope as a way into a particular set of practical questions about how we imagine and take care of futures, our own and others. Tracking some of the many modes of imagining and enacting futures that local people and the crows are taking up, the chapter offers an understanding of hope as an ecological and worldly proposition, crafted in and through specific webs of understanding and relating that enable possibilities to take root and perhaps even thrive in the world.

EXPERIMENTING

Crows experiment. From new sources of food and new sites for nesting to entirely novel modes of interspecies relating, it seems that crows are accustomed to trying things out. This is true all over the world, but in recent decades the crows of Japan have offered us a couple of particularly fascinating examples of corvid experimentation. In Tokyo, jungle crows (*Corvus macrorhynchos*) are now often considered to be a major problem for the city's human inhabitants. Taking advantage of readily available garbage, the city's crow population has grown steadily (Fackler 2008). These big-bodied, thick-billed crows occasionally swoop and injure people, and they tear open garbage bags and make a mess, but in addition to these somewhat predictable issues, they now also seem to be interrupting the city's power supply and internet access. Further afield, in the northern prefecture of Akita, crows have even been blamed for a brief shutdown of the country's high-speed train service (Fackler 2008, Fleming 2010).

Corvid experimentation is the cause of all this consternation. It seems that some of Japan's crows have taken to constructing their nests out of new materials; wire clothing hangers are particularly popular. Gathering these hangers from the environment, one after the other, crows pile and

FIGURE V.1. Illustration by Kirsty Yeomans (www.crowartist.co.uk).

interweave them into sturdy nests. The particular method that I have seen involves a crow climbing inside each clothing hanger and then wriggling, stretching, and pushing against the wire with wings, feet, and beak, to mold it into the right shape. If these nests are built in a tree, they raise no real concerns, except perhaps for the people whose clothes once occupied the hangers. But when these metal nests are built on power lines, it is a different story. Adding to these difficulties, some crows now also seem to be pulling apart cabling to use it as a nesting material, including power cabling and the city's fiber-optic network. According to a *New York Times* article, in a two-year period in the mid-2000s, "utility companies in Tokyo reported almost 1,400 cases of crows cutting fiber optic cables, apparently to use as materials for nests" (Fackler 2008).

Many strategies have been proposed and deployed by electric companies and others in Japan in an effort to mitigate these corvid impacts. Nest removal is one popular option, but crows recognize and avoid electric-company employees and rebuild when they leave (Fleming 2010).[1] Many crows have also been trapped and gassed, although it seems that mostly younger, more naïve birds get caught in this way (Lim 2010). It has even been suggested that bees might be enlisted into this "war" with crows, in this case to help reduce corvid predation on the nests of a colony of locally threatened little terns in Tokyo—the idea motivating this proposal is that bees tend to preferentially attack dark-colored animals (Ryall 2008, Cosier 2010). Here, experiment begets experiment. The destructive experimentation being undertaken by crows—which is, of course, also profoundly *creative*, depending on one's perspective—prompts humans to innovate new forms of response, to which the crows in turn respond in an ongoing cycle of adaptation. As Michiya Nakamura from the Kyushu Electric Company put it: "When we work hard to develop a strategy against them, crows develop a counterstrategy and foil it. All we can do is to look for new methods based on our experiences, apply them, and deal with the outcome" (Fleming 2010).[2]

In the town of Sendai, in northern Japan, a different species of crow, the carrion crow (*C. corone*), has found another way to make use of urban infrastructures. These birds have developed an innovative approach to opening tough walnuts: they use cars. The method is simple. It involves placing a nut in front of a car stopped at a red light and then waiting. Hopefully, the car will drive over the nut and open it (Nihei 1995). If the nut refuses to open and is simply displaced by the car, the crow will move it back. If a nut fails to come into contact with a car tire for some time, a crow might reposition it by a few centimeters to hopefully produce a better result (Nihei and Higuchi 2001). From their research, Nihei and Higuchi conclude that this behavior likely originated in the 1970s at, of all places, a driving school. From there, it slowly spread out into the surrounding region via social learning as more and more crows observed others engaging in this strange practice and tried it out for themselves. But, while cars are, without doubt, good nut openers, roads are not necessarily safe places for crows to venture. In responses, it seems that some of these birds may have refined this behavior even further. Crows in the area have now been observed dropping nuts on pedestrian crosswalks, waiting for the lights to change, and then making use of these same spaces to retrieve their tasty rewards safely (BBC 2007).

These are but two of the many instances of corvid experimentation that are taking place around the world today. Crows have likely always explored new opportunities in changing environments in this way. Humans are not a requirement; for examples, the Torresian crows (*C. orru*) who have developed new tick-removing relationships with introduced banteng (*Bos javanicus*) in Australia (see chapter 1) or the ravens (*C. corax*) who have developed hunting mutualisms with wolves, coyotes, and others (Heinrich 1999). But (some) humans and our infrastructures are today changing environments on an unprecedented scale and in a mindboggling number of different ways.[3] This is perhaps nowhere more the case than it is in urban environments. For many animal species, of course, urban spaces are simply uninhabitable. Other species find a way to eke out an existence in them (Palmer 2003, Wolch 2002). But some "synurbic species" genuinely thrive (Francis and Chadwick 2012). They thrive for different reasons (Schilthuizen 2018): some do so because they are the kind of intelligent, adaptive generalists for whom the city is something like a laboratory, a place that continually throws up new opportunities for those able to see and make use of them.[4]

Many species of crow are, of course, among those who thrive in this way. But while they are willing to try out new possibilities, crows are not the kinds of birds that generally just dive in. They are, in fact, generally regarded as "neophobic," that is, animals who avoid the unfamiliar.[5] Indeed, experiments with corvids have often shown them to be more cautious than other similar birds in approaching strange new items in the landscape. Certainly there is a great deal of variability here. Within the corvids, some species seem to be more neophobic than others, while within each species there is variability between age groups and individuals (presumably as a result of specific life histories). In general terms, adult crows seem to be far more neophobic than younger birds, with the latter often rushing to explore bright new objects (Brown and Jones 2016, Heinrich 1988, Stöwe et al. 2006a, Stöwe et al. 2006b). How can we reconcile this (adult) fear of the new with crows' prolific experimental engagements? How is it that so many species of crows thrive in transformed and transforming landscapes, places of abundant novelty, if they are more frightened to do so than many other birds? Wouldn't neophobia hamper their capacity to innovate and experiment and so to take advantage of new resources?

Fascinatingly, as Brown and Jones (2016) point out in their discussion of Torresian crows (*C. orru*), neophobia doesn't always work this way. Many corvids, it seems, are also deeply curious about novel objects in the environment. Their neophobia is paired with a kind of neophilia; their fear makes them cautious, but their curiosity alerts them to the possibilities that may lie within. "That looks interesting, I wonder if I can eat it? But I wonder, too, if it can eat me?" In this context, neophobia is perhaps an aid to experimentation—or at least to surviving long enough to get good at it, or to share its successes—and perhaps particularly so in highly novel environments. Crows remind us that experimentation is not something to be too readily celebrated; it is also, always, a risky business (Papadopoulos 2018, Lorimer and Driessen 2014). It is this combination of curiosity and caution that is thought to have produced the characteristic corvid "jumping-jack" movement, in which a crow slowly approaches and then jumps back, approaches and then jumps back (Heinrich 1988). With each repetition of this movement, the bird seems to be trying to learn, to see or hear more closely, sometimes incorporating a quick peck to test the status of a motionless animal that might yet be dinner . . . or a predator. This image has become a central one for me in my understanding of crows: they are jumping-jack birds, that is, birds whose lives are characterized by a productive and fascinating tension, by an ongoing encounter between caution and curiosity. It seems likely that it is precisely this conflict that makes them such successful experimenters and such prolific, even if sometimes troublesome, urban neighbors.

FIGURE 1.1. Torresian crow. *Source*: Photo by author.

Chapter One

INTERJECTING CROWS

Enacting Multispecies Communities

BRISBANE, AUSTRALIA

We got out of the car and walked toward a stand of tall eucalyptus trees. This is where it happened, Trixie told me. I had asked her to bring me here, to a large suburban park not far from her house just outside the city of Brisbane in southeastern Queensland, Australia. Roughly fifteen years earlier, Trixie had been called to this place by a distressed local resident after the discovery of a large group of dead and dying Torresian crows (*Corvus orru*). Most were already dead when Trixie arrived, but some struggled and thrashed around in their final throes. Their bodies were spread out around the area. Recalling the eerie and frantic scene, Trixie described how other crows crowded the tree branches above the dead, cawing loudly, visibly panicked and distressed. Crows undoubtedly notice the deaths of others of their kind; they attach meaning and significance to these events and seem to grieve these losses in their own corvid ways. These kinds of corvid funerals—or perhaps, wakes—have now been widely reported around the world.[1] In all, there had been twenty-five dead crows that day. All, it seems, had been poisoned. Who poisoned them, and why, was and still remains a mystery. Trixie, however, suspects that the culprit was likely a nearby resident, frustrated with the birds for one reason or another, perhaps their mess or their early morning noise.

My guide that day, Trixie Benbrook, is a woman with a passion for crows. In the local area she is sometimes referred to as the "crow lady" or even just the "old crow"—both names she has taken on with pride. Her entanglements with crows are multifaceted, animating and shaping her life in a range of different ways. She is a crow artist, a crow activist, a crow educator, and a crow carer. It was, of course, this latter role in particular that was relevant on the day of the poisonings: her extensive expertise in the rehabilitation of injured birds, especially crows. Sadly, by the time she arrived on the scene there was not much that could be done. Two crows were hanging on, though, enough to be rushed to a nearby vet. One succumbed to the poisoning there, but the other survived. Trixie took the bird home to rehabilitate in one of her backyard aviaries. After several nervous days, providing good food and rest, the crow was on the mend and then eventually ready to be released. Trixie captured the moment in one of her poems: "Early next morning / Doors to freedom opened / Without a backward glance / This phoenix flew."

Over the decades that she has been working with crows and other birds, this is the largest single poisoning incident that Trixie has encountered. But the poisoning of crows and other birds in and around the city of Brisbane is not an entirely uncommon occurrence. Alongside these dramatic and direct efforts to remove crows, there is a more widespread sense among at least some residents that there is a need for more official forms of intervention in crow lives, for crows to be eradicated, culled, "thinned out," or perhaps just relocated somewhere else.[2] This chapter is an exploration of the multispecies communities that are being imagined, advocated for, and enacted here. The poisoning of crows violently demonstrates the high-stakes nature of the question of community. At its most fundamental level, community is a question of how we are to be with others. More than mere occurrence alongside one another, community asks after the origins, the patterns, the modes of inclusion and so exclusion, and perhaps even the principles and procedures that structure collective forms of life. While community is frequently thought about as either exclusively human (in the humanities) or as explicitly excluding humans (in the ecological communities of the biological sciences), the communities that interest me in this chapter are decidedly more mixed up. As Dominique Lestel and colleagues note, there is no such thing as a "human community"; instead, we all occupy diverse multispecies communities—what they refer to as "hybrid communities"—that are ineradicably "based on shared meaning, interests and affects" (Lestel, Brunois, and Gaunet 2006, 172), not to mention flesh, fluids, and more.

The first section of this chapter focuses on some of the many ways in which Brisbane's crows are experimenting with new possibilities for life in this urban place. Each of these possibilities is, I argue, an ethicopolitical *interjection* that interrupts and proposes forms of shared life, of multispecies community. Here, and throughout this chapter, I draw in particular on the work of Donna Haraway and Isabelle Stengers, exploring the possibilities that reside within their thinking for doing multispecies community *well*. The second and third sections of the chapter turn to the way in which human/crow communities have been understood and managed by government agencies in Brisbane. These sections explore the way in which ideas about an ideal community—specifically those grounded in the notion of achieving a "natural balance"—rub up against more complex and multifaceted realities, sometimes violently. In response to the provocations raised both by corvid experimentation and by the violences of community, the fourth section of this chapter aims to outline an alternative approach to shared life centered on ongoing open-ended experiments in multispecies community: who we are, where we are going, and how we might decide. This experimental opening is not unlimited, nor is it one of paralysis and indecision. Rather, it is grounded in a "situated pluralism," an effort to attend to—to cultivate understanding and make room for—the interjections of others, human and not, in ways that are committed, accountable, and connected.

EXPERIMENTAL INTERJECTIONS

Standing under a relatively large roost of crows in an urban park in the east of Brisbane, I quickly got a sense of why some people might complain about the noise. It was five-thirty p.m. on a midwinter evening, and the sun was already mostly set. I had been told by Darryl Jones, an urban wildlife ecologist and behavioral biologist with extensive experience researching Brisbane's crows, that a group of them roosted in this area. I had arrived just before dusk to witness the nightly activities. Sure enough, the crows soon began to arrive, and with them the characteristic stream of steady cawing. As each new bird landed in the small stand of eucalyptus trees, small eruptions of noise and occasional squabbles were layered into the soundscape. At one point they all took to the air, circled wide, then returned noisily to the trees—a common occurrence during the settling-in

FIGURE 1.2. Torresian crows arriving at their nightly roost, Brisbane, Australia. *Source*: Photo by author.

activities of corvid roosts. In all, around fifty crows were present, gathered in three neighboring eucalyptus trees. Alongside these crows, lorikeets, noisy miners, and even a pair of kookaburras were also arriving for the night; each in their own way adding to the chaotic, exuberant cacophony. All of this activity, all of these birds, squeezed into a stand of trees occupying an area not more than 20 by 20 meters, an area separated from the closest houses by only a narrow, quiet road.

This situation is not at all unusual in Brisbane. A subtropical coastal city, Brisbane is a place seemingly overflowing with bright, noisy birdlife. While Torresian crows are widely dispersed, right across the top half and down the western third of the Australian continent, it is only in Brisbane, as far as we know, that they form significant communal roosts. These roosts can be as large as several hundred birds, fluctuating seasonally as adults depart during the breeding season to establish territories (Everding and Jones 2006). Unlike many other parts of the world, in which corvids commonly roost in the tens or even hundreds of thousands, in Australia large communal roosts of even a couple of hundred birds are rare (Everding and

Jones 2006, 21–22). There is no definitive answer to why Brisbane's crows have taken to communal life in this way. Darryl speculates that the abundance of reliable, year-round resources in the city might be a key part of the story—but then, as he has noted, this is also true in many other parts of the country.[3] In fact, as we will see, when we pay close attention to Torresian crows in and around Brisbane, as Darryl and his students do, we discover that this isn't the only area of life in which these birds are currently experimenting.

Contemporary human-crow relations in Brisbane are set within a relatively short history of urban cohabitation. While people and crows have long shared this continent—a fact testified to by thick corvid presences in the stories and cultures of many of Australia's Indigenous peoples (Clarke 2016)—until relatively recently these crows have not been part of the fabric of larger towns and cities. While there were no real studies of Australian corvid distribution before the latter part of the twentieth century, general opinion among ornithologists and biologists is that crows only really began to move into Australian cities in the 1950s or 1960s, and then only slowly (after, presumably, having been displaced from these areas by urban settlement).[4] This context matters. Given this relatively recent history of "arrival"—one within the span of personal recollection for many people—it is perhaps unsurprising that (relatively) large groups of crows in Brisbane are frequently treated as a sign of a species multiplying out of control, perhaps even having reached "plague proportions."[5]

I was initially drawn to Brisbane by media reports of significant complaints about crows. In the early 1990s disturbing articles began popping up in newspapers in Brisbane and surrounding towns: "Chorus of Crow Haters," "Cawing Prompts Call for Murder of Crows," "Locals Call for Culling of Crows," and "Crow Haters Seek Help."[6] In 2004, a ranger from the Queensland Parks and Wildlife Service is reported to have been receiving "up to 300 phone calls a year from people in the greater Brisbane region, wondering how to scare off . . . crows in their area."[7] In more recent years, other species have risen in prominence as key sources of wildlife complaints—magpies, ibis, flying foxes, and brush turkeys foremost among them—but crows continue to be one of Brisbane's more "problematic" urban residents.[8]

Complaints about crows vary significantly in their nature. Crow roosts can be noisy places. While this is true in the evenings, it is their morning calls that are particularly offensive to some residents. Starting just before sunrise, which is around 4 a.m. in mid-summer, crows begin to greet the day in their typically raucous ways.[9] Alongside their noise, crows are also a source of urban mess, with complaints about everything from excess shit to the common behavior of soaking hard food items in birdbaths and other standing water. Meanwhile, some Brisbane residents worry about the crows' effect on smaller songbirds whose eggs and young crows sometimes eat (a worry the available evidence, albeit quite limited, indicates that there is no real basis for).[10] In rarer cases, genuine instances of significant property damage have occurred, as, for example, when a group of crows took to stripping windshield wipers from a local car dealership, seemingly targeting the high-end luxury cars, requiring the owners to net the whole car yard at considerable expense. In addition, in recent years some territorial crows in the area have taken up a previously unknown behavior: selectively swooping at people during the breeding season.

I learned more about some of the interesting things that Brisbane's crows are up to and capable of in my discussions with Darryl and his students. Darryl is one of those rare biologists as interested—or at least almost as interested—in people as he is in other species. The urban setting of much of his work requires this kind of broad engagement. Over the years, he and various research students have spent a great deal of time studying Brisbane's crows, alongside a range of other species. It was while visiting Darryl one winter afternoon at Griffith University in Brisbane's south that I had my first real experience of being followed by a crow. I suspect like many other people who have paid "too much attention" to crows, I had been closely observed by them in the past; a few times over the years they had even moved in my direction to take a closer look. On this occasion though, we were very deliberately and openly followed for several minutes. As I walked through one of the many small bush sections of the campus with Matt Brown, a PhD student working with Darryl, two crows moved from branch to branch, or light post to light post, above us. They stayed at a reasonable distance but also kept a close eye on us. More precisely, they were watching Matt. Through his experiments on/with them, he has become a significant feature of their environment, a source of strange puzzles as well as food. As we walked Matt explained to me that the section of the campus in which he was conducting his research was divided into several crow territories. Over the years he had become more and more conscious of these divisions: as he moved from one territory to the next it was not uncommon for one pair of crows to stop following him, only to be replaced, sometimes quite quickly, by another pair from the neighboring territory—as though some sort of strange handover was taking place.

While in many experimental contexts this kind of familiarity with a test subject would be assiduously avoided—a practice certainly not always in the best interests of the research or the animals (Despret 2014)—in this case it was a deliberate and indeed essential component of Matt's work. He aimed to explore the presence of some of the many remarkable crow capacities that have been so well researched in other parts of the world—with species like the common raven (*C. corax*), New Caledonian crow (*C. moneduloides*), and American crow (*C. brachyrhynchos*)—but as yet largely unstudied in Australia's five crow species. Instead of working in a lab or aviary, however, Matt was working with free-living birds. This approach required him to develop "an extensive familiarisation process" in which the birds come to know him and understand something of the terms of the experimental arrangement he is proposing—without necessarily "trusting" him. As Matt put it: they are "extremely wary of all humans that pay them any attention" (Brown 2014).[11] Now completed, Matt's doctoral research has shown that Brisbane's crows are likely capable of many of the same complex behaviors as their more studied cousins (Brown 2016).

It is, in no small way, these kinds of capacities that have enabled many crows to thrive in the transformed environments of our current epoch, including urban and suburban landscapes. Amid growing patterns of urbanization around the world, crows are generally regarded as species that are doing well, perhaps "too well," out of these new arrangements (Marzluff et al. 2001). Like all generalizations, these pronouncements should be qualified. We might ask, as Etienne Benson (2011, 5) does, what it means for a species to "do well": surely we must question "a reduction of flourishing to mere numerical abundance and biogeographical distribution. We should hesitate to wish such a form of flourishing even on our enemies." Likewise, we must remember those corvid species that are clearly not "doing well," even in these abstract terms.[12] Nonetheless, in important ways many crows do seem to be "Anthropocene-ready." As we look around the world it is clear that many crows are making good use of their intelligent, adaptive, generalist ways to innovate new modes of life in changing environments.

Crows like those in Brisbane are, as Darryl put it to me, "living with us on their own terms. They're adapting to what we've thrown at them and coming out the other end very successfully." They are experimenting. In the broadest sense of the term, experimentation is something that all life does (and perhaps other things do too); adaptive responsiveness to a changing world can take various evolutionary, developmental, and behavioral/cultural forms—and, of course, it cuts across neat distinctions between these categories.[13] But some species, including many crow species, take up this experimental work in a more deliberate fashion: they try things out, adapt behaviors that are not working, watch one another, and pass on successful approaches.

In and around Brisbane, Torresian crows are experimenting in various ways. They are exploring new food items, from meat pies to recently arrived cane toads (*Rhinella marina*)—which they have learned to flip over before eating to avoid their poison glands (DEHP 2012). Perhaps one of the most interesting recent examples of local corvid experimentation is the way in which some crows have started constructing their nests on buildings. While crows of various species in Australia have long nested on anthropogenic structures, such as telecommunication towers, until very recently buildings had not been part of their repertoire. Crows in Brisbane have tended to nest in eucalyptus and other tall trees. But these birds are now exploring the possibilities that buildings can offer (Watson 2017). One pair of crows has even taken up residence on Darryl's building on the campus of Griffith University. From his office window, Darryl used to watch these crows, nesting in a nearby tree. During storms, he would see them clinging desperately to their trees and wonder how they would survive the turbulence. But, he explained, "now they don't need to do this anymore because they're tucked in this building, on a grate so that the rain just runs through the nest and down to the ground. It's completely solid and firm, protected from wind and rain."

This nesting change might sound insignificant, but Darryl thinks it is anything but. "They've made a big jump, and this has all kinds of implications down the track."[14] Will these often more sheltered nesting locations improve breeding success and lead to an increase in the crow population? Darryl thinks this is likely. More than this, though, this new nesting strategy represents a significant "cultural change," in Darryl's terms. As he noted: "All those surviving juveniles raised on a building will now think that this is what's normal. . . . From now on we can expect increasing numbers of crows to be nesting on buildings all over Brisbane, and it will just spread. There's no question about that." In fact, Darryl has gone so far as to predict that this will be "an evolutionary-significant phenomenon, quite possibly the very first step of a species diverging into urban versus nonurban subspecies."

One of the worrying possibilities taking form alongside these new nesting sites is the development of swooping behavior, targeting humans who are perceived to be threatening during the nesting season. While Australia's magpies (*Cracticus tibicen*) have long swooped at people, crows have not. It is not clear why they have started doing so—and perhaps only in Brisbane and the surrounding region—but as Darryl noted, "it could very well be because they've seen the magpies doing it" and have subsequently adopted and adapted the behavior for their own purposes.[15] These new nesting proximities and this swooping are perhaps not unrelated in their emergence. In an interview, Steve Noy, the managing director of Biodiversity Australia, a local environmental management company that deals with "problem" wildlife, told me that complaints about crows swooping are increasing yearly. He went on to say that, in a way, these crows are being "forced" into this behavior: "in some areas they can't go as high [with their nests] because of urbanization and land clearing. No one likes big trees in those areas, so the birds, if they need to evolve with us, they need to evolve their nesting. . . . They still might be 10, 15, 20 feet up, but it's just not at their preferred height."

As they get into lower trees, they're required to protect their nests more frequently and perhaps more vigilantly. "They feel like they need to," Steve noted. When taken together, these larger dynamics and emerging crow behaviors would seem to point to some difficult times ahead. Shortly after my last visit to his office, I learnt that Darryl was now being swooped by one of the crows nesting on his building. It seems that many of the new forms of life that crows are experimenting with here—from new foods and larger roosts to nesting on buildings and swooping at people—have one important feature in common: they are simultaneously responses to and intensifications of the increased proximity between people and crows in this particular place.

Brisbane's crows are active participants in the enactment of multispecies communities. As Lestel and colleagues (2006, 172) have noted, these are definitively not sites in which humans are in complete control: "Animals take initiatives in a hybrid community." In foregrounding these corvid experimental gestures as sites of emergent *community*, we are drawn into the question of how it is that we co-create worlds with others, the terms under which we come to imagine and enact collective life. Community is always a negotiation of sorts, always taking new forms. In this context, these kinds of experiments might be understood as ethicopolitical *interjections*, that is, actions that simultaneously *interrupt* the status quo—or perhaps just a particular vision of the world and how it is or ought to be—and *propose* something new, an alternative configuration of how we might get on together. To interject is "to throw or cast in between" (*OED*). It is a richly material-semiotic (Haraway 1997) possibility: one can interject verbally or bodily, but however it takes place it involves interrupting, getting in between what is and what might be, in an effort to reorient, to disrupt, to express or realize something different.

To treat corvid experiments as ethicopolitical interjections in this way is to insist on an understanding of the political that takes nonhumans seriously—not just as "objects" with reference to which various forms of exclusively human politics might take place but as themselves political actors whose forms of difference and differencing must be taken into account. This is a space of the political that works against a myopic focus on (human) language and the too-frequent assumption that politics is a specifically, perhaps even constitutively, human possibility (Grebowicz and Merrick 2013, 83–84; Wadiwel 2002). It is also an approach that refuses notions of politics grounded in singular, (supposedly) unifying systems, representatives, fora, or processes. Instead, there is what Dimitris Papadopoulos (2018) has referred to as a "compositional politics" at work here, grounded in diverse experimental efforts by a host of lively beings to craft "alternative forms of existence." Here, as in "prefigurative" approaches to politics, one's embodied enactment of change, of possible worlds and social relations, should be understood as itself a political articulation (Schlosberg and Coles 2016).[16] An emphasis on diverse sites of multispecies experimentation in this way is far from straightforward. While this political work is in an important sense always already taking place and so does not rely on or require anyone to acknowledge it as such, doing so might open up important possibilities. Doing so requires the cultivation of the critical skills and the observational capacities needed to make sense of others' lives and actions *as interjections*, to determine precisely what is being interrupted and what is being proposed; it requires that we *attend* to the diverse spaces of experimentation, of risky connection, out of which community unavoidably arises.

BALANCING NATURES

Right from the outset of my research on Brisbane's crows, I heard about the problematic roost in the suburb of Mount Gravatt. Located in a single, very large, eucalyptus tree surrounded by houses, this roost was, at one point, the nightly abode of a couple of hundred crows. By the time of my visit, however, the roost was gone. The tree had been chopped down, although no one that I spoke to seemed to know exactly when, why, or by whom. Discussing this roost one day with Darryl, he offered to take me to the spot. Although he had heard that the tree was now gone, he had not yet found the time to visit to see for himself. The roost had been of keen interest to him in the past: several years earlier it had been one of the main field sites for an undergraduate student researching human-crow conflict under his supervision. That day, we got into Darryl's car and made the short drive. Sure enough, there was no tree, but the remaining stump, perhaps four feet in diameter, revealed something of its massive size. As we considered the stump, a man approached us from a nearby building. We asked if he knew what had happened to the tree, and he replied that he believed the council had cut it down. He wasn't sure why, but he didn't think that the decision had had anything to do with crows. In fact, it seems that local people had a range of views about this particularly large roost. While some had complained loudly, advocating for the council to move the birds on—and one family is even said to have found the early morning noise so unbearable that they sold their house and moved away—other people, living no further from the site, seem to have barely noticed the crows at all, while others actively supported their ongoing presence.[17]

One might expect that local authorities would actively "control" crows in situations like this. This control might take the form of disrupting problematic roosts or nests or perhaps removing key trees, or it might involve translocation and even the killing of targeted individuals. While all of these things do happen on occasion—primarily conducted by private companies operating with state wildlife permits—the general tendency in official management approaches to crows in Brisbane is not to manage them directly at all. Instead, the relevant government agencies—the Brisbane City Council (BCC) and the Queensland government's Department of Environment and Heritage Protection (DEHP)—have adopted a live-and-let-live approach. Searching through these organizations' websites and documents, and in interviews with employees, the notion of "living with wildlife" emerged time and again. This is much more than a slogan: it is a philosophy and a management practice. It is an approach that aims to educate the public about the value of biodiversity. One employee of a government department explained to me that the goal of much of this work is to encourage a deeper appreciation for these species. She went on to tell me that learning a bit more about crows—their ecology, behavior, and habits—allows some people to relate to them and as a result to accept more readily their presence in the environment. While this approach won't work for everyone, her view was that for some people it really made a difference. In the official narratives, a key part of cultivating this appreciation for crows lies in emphasizing their role as "natural pest managers" whose scavenging helps clean up waste and dead animals that would otherwise spread disease or attract even less desirable creatures like rats and cockroaches (BCC 2014).

There is a particular working out of community taking place in these appeals to "live with wildlife." In contrast to the poisoning that opened this chapter—an event that violently announces an unwillingness to cohabitate—the official position is one of appreciation and, failing that, tolerance. From this perspective, crows are a native species and as such have a place in the environment, including the urban environment. On their website, the BCC explicitly warns local residents against harming crows: "The Torresian crow (*Corvus orru*) is a native Australian bird and is protected under State Wildlife Legislation (*Nature Conservation Act* 1992). It is a serious offence to harm crows" (BCC 2014). On the surface, an important act of inclusion is taking place here, one that, as we will see, differs markedly from the past official treatment of these birds. But this inclusion is not as unequivocal as it might at first seem. As both agencies frequently point out, given the right conditions crows can become "over abundant" and so a nuisance in the city. As the BCC website puts it: "the availability of food scraps at parks and public areas and from industrial waste bins allow crows to become dependent on artificial sources of food. . . . This causes their numbers to increase unnaturally."

The primary response to this situation, proposed by both agencies, is to restrict access to anthropogenic food. This involves covered, bird-proofed bins and educating the public about, and enforcing, waste-management procedures. The aim of these interventions is what the BCC calls a "natural balance." The state government deploys similar practical measures and rhetorical devices: "The essence of living in balance with crows is to better manage the things that attract them in the first place" (DEHP 2012). This notion of a natural balance is at the heart of the way in which communities are being imagined and enacted here. Yet it is far from clear what this actually means. In interviews with biologists, wildlife managers, carers, and policy makers, people generally expressed confusion when I asked them what such a balance might look like: did it refer to the crow population that would be present without any anthropogenic resources or perhaps just a state in which crows have a relatively benign impact on people and other wildlife? What was really being balanced here, and what was "natural" about it? How are we to reconcile such a notion of balance with the ongoing adaptation and experimentation of Brisbane's crows?

When looking for resources to help us think through the vision of community being articulated in this ideal of "natural balance," it might be tempting to turn to the literatures of ecology. But doing so is a risky business. Ecology is, of course, in its own ways a discipline thoroughly grounded in human exceptionalism. Rather than being "anthropocentric," many of the key debates and concepts of the field place humanity so far from the center that it is no longer even part of the picture. This absenting of "the human"—which is, of course, always only some humans—is itself born of an understanding in which "we" are inherently outside of ideal(ized) natural systems. Community, as an ecological concept, generally does not include humans; nor are we part of the forms of "balance" imagined. Instead, humans tend to figure in these discussions as external interlopers. As a result, these approaches can tell us very little about what a balanced *multispecies* community—composed of humans, crows, and others in an urban setting—might actually look like.

But there are, nonetheless, important lessons to be learned from these literatures, chief among them the fact that "balance" probably isn't the right thing to be looking for in the first place. Fascinatingly, the history of ecological thought in this area is intimately entangled with the question of what will count as a "community." While the search for a "natural balance" in Brisbane is undoubtedly complicated by its urban setting, beyond the city too—even in the places that we once called wilderness—the balance of nature has become a deeply fraught concept (Kirksey 2015, Kricher 2009). The Clements/Gleason debate of the early twentieth century is an important touchstone here. At its core, this debate centered on how to understand the assemblage of populations of organisms found in a given locale, now often referred to in biology as a "community." In 1916, Frederic Clements presented a theory of self-organized "climax communities": these were largely stable and tightly balanced under ideal conditions to such an extent that they might even be thought about as something like "super-organisms" (Clements 1916, Tansey 1935). In contrast, in an important article from 1926, Henry Gleason insisted that things were much more dynamic and unpredictable: ecological communities were little more than loose associations of individuals brought together by chance and shaped by an array of variables themselves subject to ongoing change. For Gleason, no predetermined—or even predictable (over a long enough time period)—climax state could be assumed (Gleason 1926, Worster 1990, Bowker 2000).

In the intervening decades, ecologists have moved further and further away from the traditional view of balance and stasis embodied in Clements's understanding. Instead, ecosystems are seen to arise and be shaped by an incredibly diverse range of factors, including climate and geology. Even the historical sequence of arrival of new species in an area is now understood to play a long-lasting role in shaping an ecological community (Fukami 2015). While there is no predictable and balanced state in such a community, there are forms of stability, equilibrium, and resilience. But they exist in the multiple, each community shaped by its own particular circumstances and likely to be thrown, or nudged, into a new state by any number of possible changes. It is for this reason that many community ecologists have today given up on the search for a "model" to explain community structure and development, instead embracing a more descriptive, cataloguing approach (Simberloff 2004) that might—for those who really like models—be thought about as an "atlas of models" (Roughgarden 2009, 524).[18] Despite this general shift in ecological theory over the past half-century, appeals to more static and balanced natures remain common both in popular discourse and in the sciences (Kirksey 2015; Wu and Loucks 1995, 441; Kricher 2009).

What might an ecology grounded in an understanding of communities as dynamic and ongoing states of affairs offer to our thinking through of the possibilities for shared life with Brisbane's crows? I found a key part of the answer to this question in a discussion with Darryl about his own ongoing interactions with the crows nesting on his building and the compromise that had been reached with these birds. He (and the several other people being swooped at as they entered and exited via the exposed stairwell) had started taking the longer, more covered route in and out of the building. But why had these birds taken particular offense to just a small group of people? Darryl explained to me that while their new nesting site protected them from the elements, it did thrust these crows into closer proximity with people: most specifically, with the man whose office window they had nested outside. While the crows were not disturbed enough by his presence to move on, it seems that they did consider him to be a threat of some sort. All of the people being swooped at were, like the occupant of this office, tall thin males.

This kind of response to "problematic" wildlife is, Darryl explained to me, increasingly becoming the norm in urban ecology. In place of a one-size-fits-all solution, we are being asked to craft locally specific responses, ones in which people are required to learn to better understand and adapt to the needs of others.[19] As Darryl pointed out in our conversation, this kind of a response is often particularly suited to living with crows. Not only are crows frequently doing new things, but when they are causing difficulties for people they are also often hard to catch and relocate, as they avoid traps: "You can catch magpies easily, they're just a synch to catch, even the aggressive ones. You're not going to be able to catch crows. . . . You might catch them once, but you certainly won't catch them again. And you won't catch any other crows that were in the vicinity and saw that one getting caught."

Where once difficult urban wildlife would simply have been shot or poisoned, this approach has become increasingly unpopular with many local (human) communities—and has always been a reasonably difficult thing to do with wily crows (see chapters 3 and 4). Darryl believes that a new kind of urban ecology is taking form here, one in which biologists will need to rethink humans as very much a part of the larger multispecies community, as he put it, as part of the household, the *oikos*, that ecology has taken as its focus but generally excluded humans from. This is an urban ecology grounded in the particularity of each individual site: "It's going to have to be much more 'custom made.' . . . It'll be local people having to adapt to the local situation." We cannot, Darryl went on to say, even be thinking at the species level: we will need instead to be asking about what individual birds are up to, how they are making meaning, interpreting and responding to a dynamic world. As Anna Tsing (2015, 23) has noted: "For living things, species identities are a place to begin, but they are not enough: ways of being are emergent effects of encounters."

Clearly, there is no room for singular visions of community here, whether grounded in a balanced nature or any other ideal. Multispecies communities will take diverse forms in response to the specificities of time and place: an atlas of possibilities. Creating broad public understanding and enthusiasm for these kinds of experimental engagements between people and other species must be a key part of the work of this new urban ecology. It is a task Darryl frequently takes up: through various media outlets, through popular writing (Jones 2018), and through his own research collaborations, he is cultivating understanding and making new kinds of room for wildlife in Brisbane.

To a large extent, this is also precisely the kind of approach to public education and acceptance that the BCC is working toward with their emphasis on "living with crows": asking people to be accommodating, to make room, to learn to appreciate their corvid neighbors in new ways. But things are not quite this simple. As we have seen, this approach is also framed as one that should conform to a "natural balance." While it is far from clear what such a state might look like in concrete terms, it nonetheless represents an important effort to define the who, how, and why of shared life. In this way it does powerful, even if not always easily recognizable, work: it frames and curtails the kinds of possibilities we can imagine and enact with crows. I suspect, in fact, that the primary function this framing plays is in providing the appearance of precisely this kind of restriction: reassuring the wider community that there is a goal, that there are limits. In doing so, however, this framing also opens up space for important forms of violence and exclusion.

VIOLENCES OF COMMUNITY

The crows of Brisbane are no strangers to the balance of nature. This is by no means the first time that this balance has been invoked to imagine ideal forms of cohabitation with them. As early as 1892, roughly seventy years after the founding of a permanent British settlement in the area (on what was Jagera Indigenous land), members of the newly formed Natural History Society on an excursion to nearby Stradbroke Island remarked on the abundance of crows, "doubtless attracted by the garbage from the settlement."[20] Twenty years later, crows were still very rare in the city of Brisbane proper, but farmers in the surrounding regions pushed through the formation of a Flying Fox and Crow Destruction Board that paid bounties on dead animals. Predictably, this approach was grounded in the view that these "pest" species damaged crops and killed livestock. But underlying such claims, at least for some, was the further view that crows—in their numbers and perhaps also their behavior—had become something different to what they once were: unnatural, out of balance. Spurred on by greater food availability—whether in the form of garbage or agriculture—they were becoming a problem. As one letter to the editor of the *Queensland Times* put it in 1912, the year of the formation of the Destruction Board, "the crow of 50 years ago is not the crow of to-day." The author goes on to explain that

> as soon as [post-1788 colonizing] population came into the country the crow apprenticed all his youngsters to the butchering line, and went in for the lamb and chicken business. . . . When it is planting time for oats and wheat his lordship is always on the job to sample the quality. The crow is so educated now that if he saw the crow of 50 years ago he would have a petition to the Government to have him expelled to the Fiji Islands.[21]

Crows are here drawn into the grand narrative of "modern" agricultural development (Muir 2014; O'Gorman 2012; Swanson, Lien, and Ween 2018). This is a narrative that pits the supposedly civilized against the supposedly primitive, in the name of a broader colonizing project, one in which the worlds of diverse humans and nonhumans have long been "sentenced to disappearance in the name of the common goods of progress, civilization, development, and liberal inclusion" (Blaser and de la Cadena

2018, 3). The racism is palpable in this quote, as is the disdain for crows. The writer concludes by noting that he is not exactly supporting their complete eradication: "I would like to see . . . two crows in a cage, and all the rest dead."

But the "balance of nature" was also put to work in the opposite direction in this period. Far from being out of balance and in need of control, crows were positioned by some as vital to keeping nature in proper order. To a meeting of the Destruction Board in 1919, A. H. Chisholm, the president of the Royal Australian Ornithologists Union, argued against the killing. According to the *Telegraph*:

> While admitting that the crow and kurrawong did a good deal of damage in various ways, he held that these two native birds had their own place in the balance of nature, and had much to their credit in the killing of noxious insects. . . . Scientific evidence on the subject was quoted, and the opinion offered that it was a very dangerous thing to attempt to exterminate any native bird whatever. This consideration, according to the lecturer, did not hold good in respect of the starling and sparrow. Neither of these birds, he said, should ever have been introduced to Australia.[22]

Over the following decades, hundreds of thousands of crows and flying foxes were killed for the bounties on their heads (along with other bird species later added to the Destruction Board's mandate). When reading newspaper articles from this period, a single overriding theme emerges. Certainly, a diversity of opinions is expressed: as one commentator succinctly put it, "the black marauder of the air has his friends and his enemies."[23] But shared across all of these points of view is an effort to evaluate whether, on balance, crows represented a positive or a negative influence *for the farmer* (their key potential positive was in the control of blowfly infestations among sheep). While the occasional voice—like that of Chisholm—called for a broadening of focus beyond the farm to take into account their impact on "flora and fauna" also, such positions always needed to be grounded in a primary appeal to agricultural utility. Within the context of the thoroughly productivist mindset that dominated this period—what Deborah Bird Rose has called a monological vision (2008)—the fate of crows and others could legitimately be tied to the answer to this question (O'Gorman and van Dooren 2016, O'Gorman 2014).[24]

The intervening decades have seen an important shift in the local treatment of crows. The Destruction Board was wound up, and like all other native species Torresian crows are now protected in Queensland. This protection is far from absolute—as is dramatically demonstrated by the current persecution of flying foxes (Rose 2011a, 2013a)—but certainly bounties are no longer paid and permits are required to "take" native wildlife. As we have seen, in many ways the kind of view outlined by Chisholm now dominates much of the government rhetoric, albeit couched in a different language. As the BCC (2014) puts it: "the crow plays an important role in natural pest management. . . . Crows also remove road-killed animals by feeding on carrion, and disperse many native vegetation seeds through defecation. The long-term conservation of this species is necessary for maintaining biodiversity." In short, the crow is now often positioned as a valued and integral part of a larger community of life. While people may not always like sharing space with them, they are called upon to tolerate, perhaps even to appreciate, crows for the role they play in achieving a desirable, balanced environment.

On the surface, then, it might seem that while the "balance of nature" can be made to do violent work it can equally be utilized to challenge that same violence. This kind of ambiguity is no real surprise: it has long been noted that nature and the natural can be, and are, strategically deployed (Williams 1980, 70). But there is something about the way that the balance of nature is invoked in contexts like this that is somehow less malleable. There is, it seems, a fundamental, even if not always readily apparent, violence inherent in *all* such framings. In this regard—while certainly preferable in *many* ways to both destruction boards and vigilante poisonings—the positions offered by Chisholm and the BCC never quite escape this space of violence. Instead, they simply reframe and reposition it. Again, the logic of community is the key to understanding how and why this is the case.

A prominent strain of philosophical and political thought, beginning perhaps with the work of Jean-Luc Nancy in the early 1980s (Nancy 1991, James 2010), has sought to rethink traditional notions of (exclusively human) community grounded in a common identity or purpose: community as that which is held in common (Devadas and Mummery 2007). Whether centered on forms of nationalism, religion, a shared tradition and history, or indeed a shared fate, according to these traditional understandings community is a more or less unified collectivity grounded in affiliation or identity. Understood in this way, community is often positioned in opposition to nihilism: as that which provides meaning and purpose, holding purposelessness at bay (Esposito 2009). In contrast, Nancy and others have explored the "double violence" (Morin 2006)—an immunitary violence (Esposito 2010)—inherent in efforts to define and shore up this kind of collective life. This is a violence that operates both externally and internally: through the exclusion of those that do not fit, that do not share what ought to be common, as well as through the ongoing regulation of those who are included but who nonetheless threaten the collective by failing to conform adequately or permanently (and so must be altered, cast out, or killed). These philosophers, each in their own way, have developed alternative understandings of community as a kind of "unworking" that seeks to interrupt closure, self-certainty, and exclusion (Morin 2006, James 2010).

The community of natural balance—like any other utopic community lost or yet to come—contains within it the seed of this double violence. The most obvious violence at stake here is that of *exclusion*. From this perspective, while crows may now fit within the imagined community, others have not been so lucky: for A. H. Chisholm, the starling and sparrow—neither of which "should ever have been introduced to Australia"—are not afforded this same consideration. Their removal will not disturb anything; rather it will help set things right. Similarly, in contemporary Queensland, while these crows and other native species are now protected as parts of a cherished biota, non-natives tend to be vilified and persecuted diligently.[25] In fact, it was the very act of inclusion that protected Torresian crows, grounded in a new sense of a valued *native* community of biodiversity enshrined in state and commonwealth legislation, that simultaneously created an intensified zone of exclusion, a space of abjection occupied by all of those now positioned as non-native (many of whom were deliberately introduced to Australia by settler colonists and protected not that much earlier). As such, the globetrotting house crows (*Corvus splendens*) that arrive in Queensland from time to time as stowaways on cargo ships—like their brethren I will discuss in chapter 3—are subject to immediate "destruction" (Csurhes 2016). In short, while the state no longer sanctions violence against native crows, the particular enactment of community that positions them as valued life simultaneously creates a "constitutive outside" (Butler 1993, 8) populated by countless others. This violence lies at the heart of—it is in fact *foundational* to—this imagined community of natural balance. To analyze this dynamic of exclusion is not to offer any general conclusion about whether "introduced" species ought to be accommodated in new environs, nor is it to dismiss the significant effects they sometimes have. It is, however, worth noting that black-and-white framings that valorize the maintenance of pure "native community" are not the best way of grappling with the complexity of these issues (Davis et al. 2011).

But there is a second, more subtle form of violence at work here: the internal violence of *conformity* with pattern or identity. Those included within the community can only retain their position by conforming to its dictates, in this case by staying in balance. Being "native" is not enough. As these crows and other so-called nuisance species remind us, it is still very possible, under the proper conditions, for community membership to be rescinded or made conditional, leading to anything from death to milder forms of reeducation or realignment. But beyond official measures, this positioning also colors popular opinions and responses. When the BCC declares that anthropogenic food sources can enable crow populations to "increase unnaturally" or the state government explains that restricting access to such food is the key to keeping the crow population "in balance," a dangerous logic is reproduced. From this perspective it becomes possible to view crows as unnaturally abundant, perhaps especially in those places where they are eating anthropogenic foods. Of course, crows are very often eating these kinds of foods—they are, after all, adaptive, generalist scavengers—and so any undesirable crow is able to be counted among the "too many": the ones who really ought not to be there.

This positioning enables self-righteous attitudes among those who dislike or are disturbed by crows. In the most extreme cases, this rhetorical framing might even allow people to feel justified in killing and other acts of persecution that can be viewed as managing an unnatural population, restoring things to a balanced state. Again and again we encounter precisely this logic in Brisbane. A particularly clear example of it can be found in an article in the *Courier Mail* from the height of the period of crow hating in the late 1990s. Here, a representative of an organization called Australian Native Animal Care is quoted as saying: "Brisbane's burgeoning crow population was 'beyond a joke.' 'It's incredible. They really are out of proportion compared to places where nature's completely in balance.' . . . [He added that] humane culling of the city's crows would make environmental sense by relieving the pressure on other native birds."[26]

We see here the way in which "balance" allows care for (some) native animals to become a practice of killing (other) native animals. This practice of care is made even more problematic by the fact that there is no real evidence for the view that crows are detrimental to other local bird populations. The fact that this position can so readily be taken up by a wildlife carer is both fascinating and disturbing but perhaps not surprising. As Laura McLauchlan (2018) has documented so well, wildlife care is always a partial and compromised work, one demanding critical attention to its own constitutive violences.[27]

In my conversations and research I frequently heard about "crow plagues" and even "rogue crows"[28]—always as part of an account of a species out of balance that is now affecting local people and wildlife and in need of "thinning out" for the good of all. Most of the time these were simply calls for someone else to remove or kill crows. But, as we have seen, in a few cases people take things into their own hands. We lack any concrete data on the number of poisonings that occur, let alone on why they occur. In many cases autopsies are not conducted. When they are, they are often inconclusive. Even in cases where poisoning is likely, who carried it out and why usually remains a mystery. I am not aware of any person being prosecuted for poisoning crows or other wildlife in Brisbane. But *accusations* of poisoning are relatively common. Inside all of this uncertainty, anecdotal evidence points to one strong pattern. Almost all of the people I interviewed about bird poisonings in Brisbane—from wildlife carers to veterinarians and government managers—noted that these events seem commonly to occur in the context of one person being annoyed by a neighbor who feeds birds: annoyed, that is, by the establishment of a very visual and vocal, an explicitly "unnatural," abundance—and with it "excessive" noise and shit.

What is at stake here is shifting and elusive. While the official rhetoric and much of the practice points toward a live-and-let-live approach to crows, beneath the surface things are more ambivalent. Various forms of violence are tangled up here with the work of achieving a natural balance: given legitimacy, enabled, made obvious and routine, made to *matter*. Violence is, of course, an inescapable part of shared (multispecies) life, no matter how community is imagined and enacted. "Living with wildlife" is always only living with some wildlife, and it is always at the expense of others. But this particular way of doing community effaces and naturalizes its foundational violences. These violences are integral to the way that concepts of both "community" and "nature" are often deployed: both terms claim to name an empirical reality and at the same time an ethical or political good. And so, the "balanced nature" envisaged here is taken to be both a specific, even if nebulous and imprecise, *state* that might be achieved or at least aimed for and, at the same time, a *justification* for that state, an incontestable ideal.[29] In this context, the question becomes: is it possible to inhabit this space, to constitute community, otherwise? Might we hold open more room for an ongoing questioning—for testing, for play, for the always non-innocent work of experimenting—with what else might be possible?

EXPERIMENTS IN COMMUNITY

The second time I met Trixie she had invited me over to her house to see the many aviaries in which she keeps injured birds, the aviaries that had been a temporary home to the one surviving crow from the mass poisoning that opens and animates this chapter. On the day of my visit, the ten or so aviaries in her backyard were occupied by a range of birds, including magpies, lorikeets, magpie-larks, figbirds, and tawnie frogmouths, as well as a single crow that had been hit by a car. As Trixie showed me around, we talked about the work of caring for birds, about her political advocacy on behalf of crows, about her corvid poetry and photography, and especially about her work as an educator.

Until that day, I must admit, I had never even thought about how one might go about making a bird's nest. In our discussion, however, I learned that constructing a crow's nest from scratch is one of the exercises Trixie employs in her wildlife classes with children. Using an actual crow's nest as a model—one that turned up in a box many years ago on her doorstep, complete with orphaned crow chicks in need of care—she invites children to collect scattered twigs and sticks and make their own nests. The catch is that they can only use the thumb and index finger of one hand: a hominid approximation of a bird's beak. Collecting one stick at a time, back and forth they go. Other children play the parts of (less lethal) cats and dogs, obstructing them in their work. Once their sticks are assembled in a pile, the children discover that weaving them together into anything resembling a nest—again using only their "beaks"—is far more complex than it seemed at first. Vinciane Despret (2004) might call this an

FIGURE 1.3. The crow nest used by Trixie in her classes. *Source*: Photo by author.

experimental anthropo-zoo-genesis: helping children in a modest and thoroughly imperfect way to enter into the worlds of crows. All of this, in a fun and memorable way, aims to educate students into a new appreciation of crows and their nests.

Like crows trying out new urban nesting sites, I understand Trixie's art/labor of nest making as an experiment in community. This is an experiment grounded in practices of attentiveness: of paying *attention* to another as a means of learning to *attend* better to what matters to them. Attentiveness is an ethical work in the thick sense of this term, a work of worlding well with others. In asking children to be attentive to crows, Trixie cultivates space for a particular kind of connection and with it a particular kind of community. She cultivates space for the kind of openness to difference that might enable us to appreciate better our own partiality and in so doing be transformed by it. I take this exercise as a reminder that community is not a collectivity composed of atomized, preexisting individuals; rather, it is an ongoing process in which all of us are at stake. The work of community is not one of "preconstituted identities" but rather a process of "constituting those identities themselves in a precarious and always vulnerable terrain" (Mouffe 1999, 753). All of Trixie's work as a poet, a filmmaker, a photographer, and an educator might be understood in the same light: as an effort to experiment, to interject, and in so doing to enact possibilities for flourishing modes of shared life.

Trixie's tireless and diverse efforts on behalf of crows have been a major inspiration in my efforts to think through the difficult work of crafting flourishing multispecies communities. This is not simply a question of dwelling with others—of recognizing that multispecies community is an inescapable fact of shared earthly existence—but rather one of taking up the ongoing *work* of making communities well. In the final section of this chapter, I offer my own experimental interjection, my own proposal that might interrupt the status quo. This proposal is guided by the diverse experimental gestures taking place in Brisbane—those of Trixie, of Darryl and Matt, those of the crows themselves—and aims to respond in some way to the challenges raised in the first three sections of this chapter. These challenges are grounded simultaneously in the empirical reality of communities—that they simply will not conform to fixed and static forms—and in a normative imperative that the endeavor to have them conform is unavoidably violent in ways too often taken for granted or effaced (Grebowicz and Merrick 2013, 93).[30]

My experimental interjection takes the form of an alternative notion of community, an understanding in which community is a never-to-be-finished project, an ongoing crafting of lives in common, conducted as a work of *situated pluralism*. From this perspective, it is not enough simply to replace natural balance with another singular goal, whether one dreamed up by conservationists, philosophers, or some sort of local coalition. The fundamental task here is not to question this particular vision of an ideal community but to raise as a question the possibility of any universal goal: to hold permanently open the question of community, to refuse the assumption that we know, that we *can* know—whoever "we" might be—the who, how, and why of shared multispecies lives. Such a work of community is, deliberately, "incapable of giving a 'good' definition of the procedures that allow us to achieve the 'good' definition of a 'good' common world" (Stengers 2005, 995). In place of a singular goal, a singular "good" of any kind, it is a work of *multiplication*: crafting spaces for the collaborative imagination and ongoing enactment of diverse possibilities, an open-ended work toward somethings better.

It is in this context that I understand the work of community as one of situated pluralism. This is a definitively active and constructive work, even if it in some ways shares the concerns that animate Nancy's and other's efforts to "unwork" community. It is an effort to hold open room for diverse worldings—for a diversity of modes of knowing and being, of approaches to what shared life might consist in and how it ought to be achieved. This is an already existing diversity, but it is one always also taking new form, emerging in new ways, through dynamic experimental processes of intra-activity. But to hold open room for difference in a "situated" (Haraway 1991) way is explicitly *not* an effort to practice an unlimited acceptance of any and all possibilities. Inside worlds in the making, there is no room for an ethics/politics that floats above the fray, insisting only on the metaprinciple of openness and inclusiveness: a bland priesthood of disengaged purity.[31] To insist on an approach of absolute openness is to fail to take responsibility for the fact that worlds—and communities—don't work like this, that not all possibilities (ideals, perspectives, projects, even bodies) are compatible, capable of mutual flourishing in the same placetimes. As Mario Blaser (2016, 563) notes: "Sometimes different worldings may coexist—enabling each other or without noticing each other—but at other times they interrupt each other."

A situated perspective is one that refuses the lures of both a singular outcome and boundless openness to, or inclusion of, everything. In their failures to situate themselves, both of these approaches are unable to yield a *responsible* positioning. To be situated is to be in the world, partial and finite; more than this, it is to insist that this is the only way in which one can be (Haraway 1991). From this perspective, living well with multiplicity must take the form of an ongoing effort to become accountable for ones' own worldly positionality, acknowledging the necessary partiality—the limitations and even violences—of that position. The goal of such an acknowledgment cannot be to escape positionality but rather for it to become an invitation into *connection*, into the work of crafting better understandings and ultimately worlds with others. The never-to-be-finished result is not a singular "correct" or "ideal" outcome but something much more kaleidoscopic, networked, partial, ongoing (Haraway 1991, 590). These are communities made from the inside, in the multiple, again and again, with others. The challenge is to inhabit this space *well*, in a way that is connected, accountable, and committed.[32]

Doing so requires a pluralist impulse. Pluralism is not relativism; it is not about the equal validity of all positions. Rather, it is an effort to inhabit a world from ones' own situated perspective *but* to do so under the weight of a general openness toward, a respect for, difference. This requires taking up something like what William Connolly has called a "bicameral orientation."[33] The real challenge arrives when the worlds we are working to enact collide. Here, the pluralist impulse demands that we take up our worlding work in ways that remain *as open as possible* to others' lives, projects, and possibilities, indeed, in ways that are open to being transformed by others. This is not about openness in the abstract but rather openness to possibilities for accountable modes of *connection*, of experimenting and interjecting-with. In this way, situated pluralism demands much more than the mere "tolerance" of others, in which preexisting bodies agree to disagree. The effort to understand and make room for difference carries the profound possibility that everyone might be *redone*: that openness to another might enable other possibilities for either or both parties (Cheney and Weston 1999). Like children making crows' nests—learning, however imperfectly, to appreciate the difficulty, the fragility, of another's life and world—we, or they, might be transformed by this openness, openness not to "the Other" in the philosophical abstract but rather to "an other" or others (Grebowicz and Merrick 2013, 103).

Situated pluralism is grounded in neither a commitment to consensus nor to ongoing agonistic contest (Mouffe 1999). Neither of these approaches to difference and differencing—squashed into the singular or held permanently apart—can contain diverse possibilities for experimental modes of life, of learning to connect and interject-with and -against others. Subjects and communities in-the-making are always becoming otherwise, their forms of difference pulled into new alignment and/or dissensus. To situate oneself, to attend to and take responsibility for positionality and its possibilities for connection, is necessarily to simultaneously craft forms of interwoven *kinship and difference* with multiple others (Plumwood 1993).

To take up the work of crafting community well with the crows of Brisbane is not, therefore, about embracing an open-ended acceptance of their populations and of any and all corvid projects. Nor is it to devise a singular approach to how human-crow relations ought to be valued, understood, and conducted. Rather, it is about multiplying spaces for experimentation with/for a diverse range of peoples, of crows, of countless others, asking *what else might be possible together? How might we be redone by one another, at stake in one another?* Taking up these questions requires us to attend to the kinds of corvid experiments discussed in the first part of this chapter. When crows swoop or construct their nests in new places, they are interjecting in the world in a way that is thickly material-semiotic, simultaneously bringing about possibilities for shared life and communicating something for those able to understand. When corvids are viewed in this way, we might move away from an understanding of these emerging practices as altered behaviors that give rise to problems, or violent reprisals, or even acts of tolerance. Instead, or in addition, they might begin to appear as opportunities to explore possibilities for multispecies community in which the terms of life are coproduced rather than imposed (also see chapter 4).

A swooping crow in Brisbane would seem to communicate the experience of a threat and propose an alternative mode of cohabitation. The magpies, from whom they perhaps adapted the behavior, often begin with a couple of warning swoops, asking a would-be intruder to turn around or demonstrate in some other way that they are not a threat, exploring whether they might be "educable to the standard of a polite guest" (Haraway 2008, 24). It is only if these warning swoops fail that magpies adopt a more direct approach, making physical contact with their sharp beaks. As Gisela Kaplan, perhaps the foremost expert on the Australian magpie has put it: "magpies are extremely polite and formal birds—they have real etiquettes and breaking them makes them frustrated, confused and sometimes angry."[34] At this stage at least, Brisbane's crows seem to be doing things a little differently. Darryl believes that they are using only "warning swoops" (that is, not making physical contact) and relying on this gesture to make their intentions known. The effectiveness of such an approach remains to be seen; it rests in large part on the capacity of people to interpret meaningfully and respond appropriately to the communication.[35]

While the emergence and transformation of community goes on irrespective, efforts to craft community *well* are circumscribed by our—all of our—abilities to understand, imagine, and render legible and sensible others' interjections. Interpretation is never straightforward. These are sites full of the uncertainty that always surrounds questions of meaning and intention in contact zones (Pratt 2003, Haraway 2008). Outside the illusion of complete transparency to one another, we are left to "make" sense *with*, not for, others (Haraway 1992). While there are no "official" ventriloquists who might act as "representatives" or "spokespeople" (Kirksey 2012), there are rich and diverse sets of resources for making sense and contesting meaning with and about crows, including the knowledges of biologists, of wildlife carers, and many others. It is for this reason that the work being carried out by people like Trixie and Darryl is so important. In different ways, each of them cultivates spaces for better understanding and cohabitating with Brisbane's crows.

The approach to "living with wildlife" adopted by the BCC is also, in many ways, a hopeful one. As we have seen, at its core this is an approach grounded in openness to corvid difference, aiming to encourage local people to better understand and make room for crows. It also places a central emphasis on people's own responsibility for crow "problems" and asks them to address their own behaviors as root causes. Ultimately, this approach articulates a vision in which the urban environment, the multispecies city, must be a compromise between a diversity of ways of being. At the same time, practical efforts to crow-proof bins and reduce access to anthropogenic food might help defuse some potential sites of conflict.[36] This is, without doubt, one of the more positive "official" approaches to human-crow relationships that I have encountered in the course of researching this book. Yet, as we have also seen, it has important limitations. The forms of "living with" others that can emerge inside a space of "natural balance" are curtailed in critical ways, limited—from the outset—to specific kinds of others and specific modes of abundance and lively expression. The problem with the limitations imposed by natural balance is not that any limitations at all are to be avoided—they are an inevitable part of all concrete articulations of/for worlds—but rather that these particular limits are predetermined, one-sided, self-justifying, and fixed, while also being vague and traveling with the normative weight of "nature." An opportunity is missed here to do more interesting and more experimental work. Such an approach might begin from the conviction that we don't know what forms this multispecies community should take, we don't know what will become of human-crow relationships into the future, and that this is as it should be. In this context, "living with crows" would require us to attend to them carefully, to hold open space for their experimental gestures wherever we can, to continually (re)imagine and propose new possibilities for flourishing multispecies life.

While our worlds will always, and must always, be "partially shared" (Haraway 1991)—and we can certainly work to do this sharing in better and worse ways—there can be no final settlement or end to the composition; nor should we hope for one. Instead, a situated pluralism can offer only the ethicopolitical promise of paying attention in necessarily partial, ongoing, multiplicitous ways—without the possibility of any pregiven or, indeed, ever-to-arrive authorized set of principles or procedures for doing so. Paying attention as the basis of efforts to make connections and craft new possibilities together: this might take the form of nest making with children, of researching and communicating about crow behaviors and their urban propensities, even of avoiding an exposed stairwell for a few months each year.

In Australia and around the world, many other possibilities are emerging. The Northwestern crows (*Corvus caurinus*) of Vancouver now have an interactive map showing where they're swooping (MacKenzie 2016). Magpies in Australia frequently have similar programs. Perhaps Brisbane's crows will soon join them? Or perhaps we will find yet other alternatives? In place of the "fake eyes and spiky helmets" that people sometimes wear in Australia to deter magpies from swooping—creative tools and tactics for the kind of diplomacy that will be discussed in detail in chapter 4—Gisela Kaplan recommends an approach that is both simpler and more demanding: that people endeavor to "make friends" with local magpies (Brown 2017). Magpies, like crows, are very good at recognizing individual humans and remembering both friends and foes, which, of course, doesn't mean that they don't sometimes make mistakes. By establishing oneself as a friend—which often but by no means always involves the provision of food or water—magpies are much less likely to feel threatened and swoop during the tense breeding season.[37] Might crows respond in similar ways?

Diverse, ongoing changes will be needed to enact flourishing human-corvid relations in urban places, changes to understandings, to modes of relating, as well as to infrastructures (Seymour and Wolch 2009). But these challenges cannot all simply be outsourced to centralized wildlife-management interventions. A situated pluralism requires that we actively take up the shared task of attending, of experimenting together, in the countless, specific sites in which living beings meet to propose possibilities for shared life. This is a vision of multispecies community that draws in significant ways on the insights of the natural sciences—especially those of scientists like Darryl who are attending carefully and crafting "custom-made" responses—but that ultimately insists that it is not up to this or any other group of "experts" to decide what forms shared lives should take. These are fundamentally ethicopolitical questions—which is not to say that they are not simultaneously scientific questions—and they must be the ongoing product of a collective working. Part of this work must necessarily involve critical efforts to understand and intervene in the larger dynamics—the histories, the webs of relating, valuing, and understandings—that are producing our current reality. In Brisbane, as previously noted, ongoing habitat loss and urban development are likely a key part of this story, thrusting crows and people into new proximities and contestations. How might these processes be interrupted *with* crows and others? How might alternatives be proposed that incorporate us all from the outset?[38]

Within this broad space of experimental interaction, taking up an approach of situated pluralism means holding open room for others in a way that is committed, connected, and accountable. This is by no means something that all of the beings engaged in experimenting and proposing their own agendas for multispecies community are likely, or indeed able, to take up. It is one thing to insist that crows are enacting community and that the people of Brisbane might hold open room for them to do so, but can we really expect the crows themselves to take up the work of attending to and making room for others?

The beginning of an answer to this question might be found in the Cobourg Peninsula in the Northern Territory, Australia, 2,800 kilometers to the northeast of Brisbane, as the crow flies. Here, other Torresian crows—the same species found in Brisbane—have struck up an unlikely relationship with banteng (*Bos javanicus*), a species of wild cattle that only arrived in Australia in 1849, introduced to this one small area as part of a failed British settlement (Bradshaw et al. 2006). In recent years, scientists have observed crows landing on the backs of resting banteng. The banteng will then roll onto its side and lift its upper legs (which is not a comfortable or easy posture for a banteng), so that the crow can access the area under the legs and belly. Moving into this space, crows have then been observed removing ectoparasites, likely ticks, from these exposed areas. It is not known where this behavior came from. Torresian crows aren't known to have similar relationships with any other mammals, and banteng around the world aren't known to deliberately expose themselves for grooming by any other bird. Yet this is happening here. We will never know which crafty individuals struck up this mutualism, how those first awkward interactions took place, how a proposal was made, and how an agreement was reached that vulnerable proximity to each other was worth the risk. But we do know that this behavior is spreading as more and more banteng and crows around the region get in on the action (Bradshaw and White 2006).

Is this a work of situated pluralism, or might it at least be an opening into a related space of community making? At the very least, crows here seem to be adapting to the needs and differences of others, practicing attentiveness to others' worlds, learning to understand what matters to banteng, and working actively to cultivate the space for it. Of course, the action and initiative is not all on one side of the relationship, but I'd like to stick with the crows as much as possible (not least because we would need to know and say much more about banteng to explore their role in any depth). A naysayer might insist that these crows are simply making space for their own needs: benefiting from free ticks. Setting aside the fact that there are many easier ways for a crow to find food, such an objection misses the point. A situated pluralism does not require altruism. It is a practice of attending to and making room for difference. In some cases this might require altruism, in others something more like mutualism, and even parasitism is well and truly possible—although all of these terms need to be complicated and drawn out of the black-and-white frames that they too often occupy in biological discussions (van Dooren and Despret 2018). In this sense, it is about the effort to understand another and the ability and willingness to make room for them where appropriate. Not endless, unlimited room—but room where one can.

In some cases, doing so might require imaginative acts of something like what the primatologist Frans de Waal (2008) has called "targeted helping." In introducing this term, de Waal reminds us that at least some animals are well and truly capable of understanding and working toward the specific "good" of another (even if it is very different to what constitutes their own good); for example, the bonobo who climbed to the top of a tall tree and carefully unfolded the wings of a stunned bird before trying to release it back into the air (Preston and de Waal 2005, 19). Might these crows of the Cobourg Peninsula, in a similar way, simply/also be trying to help banteng out? We will, perhaps, never know, but such a possibility does not seem to be beyond the realms of possibility or indeed beyond the capacities of clever corvids (see the "Gifting" and "Cooperating" vignettes). In a similar vein, we might ask: could Brisbane's crows also be carefully attending to and deliberately making room for others? In the context of a dynamic urban environment in which a range of other species are also arriving, adapting, and thriving in new ways, this is definitely a possibility. As Darryl noted: "they're learning to live with each other. . . . Who knows whether these are positive, communicative, cooperative relationships that are being formed."[39]

Ultimately, a community grounded in practices of situated pluralism cannot overcome the unavoidable reality that not all forms of difference, not all understandings and modes of life, will be able to be accommodated. Of course, not all challenging forms of difference are other-than-human. Many are much closer to home yet no less difficult to understand and live with for this proximity. How, for example, might we attend to the poisoner whose actions opened this chapter? How might we learn to better understand and make room for a person, woken at 4 a.m. each day, driven to an act of violence? Situated pluralism does not require agreement with or even acceptance of another's position. It requires something more difficult: an effort to genuinely understand another, often across vast and perhaps even incommensurable difference. In some cases this understanding might lead to change, to greater acceptance, to possibilities for shared flourishing; in other cases it will become the basis for renewed disagreement and contestation, for a concerted stand against another. Either way, what is required is an effort to take up the work of enacting community with others without the comfort (for whom?) of any final, or singular, answers. The goal, in short, is to denaturalize community without succumbing to the paralysis of uncertainty: to refuse to allow any approach, any understanding, to become part of "the system," built into the assumptions that "we" make about who counts as a member of the community and what processes ought to guide efforts to imagine and craft shared worlds. In this way we might get on with the careful, attentive work of experimentation and interjection with one another—human, crow, and countless others, asking again and again: What might be possible here and now? What other forms might multispecies community take?

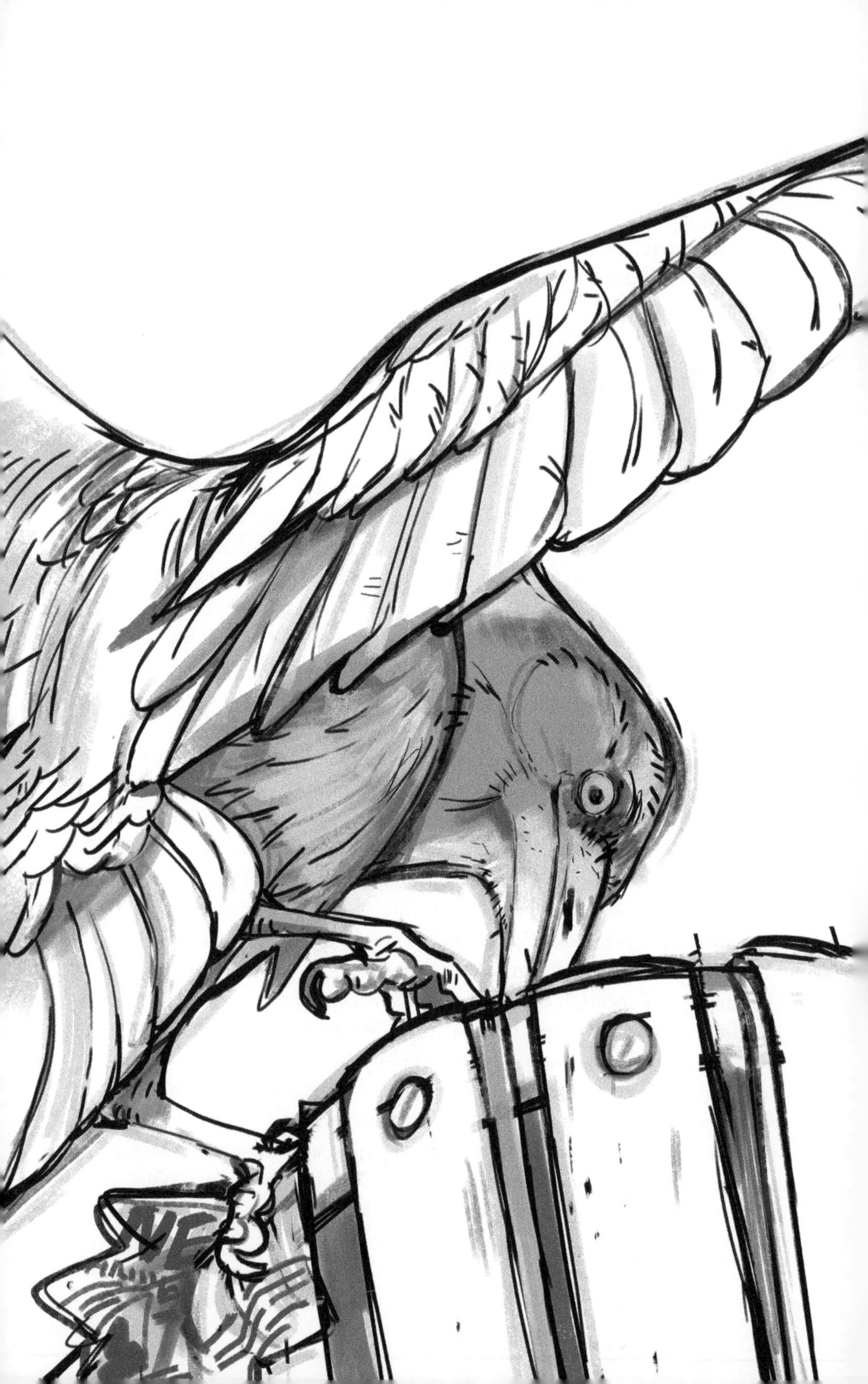

STEALING

Crows steal. In fact, according to media reports, in the early 2000s a series of corvid heists took place at the Membury service station on the M4 motorway in England. In each case, events unfolded in pretty much the same way. Two rooks (*Corvus frugilegus*) arrived and took up positions on opposite sides of the top of a garbage can. Working in tandem, they pulled the plastic liner up with their beaks, securing it at this new height with their feet before reaching down again with their beaks to pull it up further. Repeating this action about twenty times, the birds gained access to the once inaccessible waste at the bottom, bringing it ever so gradually within beak's reach (Clayton 2015b, 229). While no research has been conducted on these particular rooks, many of the remarkable behaviors that went into this heist—teamwork, patience, and calculation—have been experimentally demonstrated by corvids over the years, including the "pull-secure-pull" technique (Heinrich and Bugnyar 2005) and even the synchronized, cooperative pulling by two birds to secure a food reward (see "Cooperating").[1]

However, it isn't entirely clear that anything was really being "stolen" in this case: can one really steal what another has discarded? But there is an

FIGURE V.2. Illustration by Kirsty Yeomans (www.crowartist.co.uk).

important act of theft lingering at the edges of this action. When the garbage was finally within reach of the rooks' beaks, reports indicate that one of the birds would start tossing the food over the side of the bin while the other, or perhaps a third rook in on the action, stood guard on the ground to ensure that the hard-won food wasn't stolen by others. So the real site of potential theft took place *after* the elaborate heist, once the food had been secured. Here, in this seemingly more mundane, everyday space of encounter—crows squabbling over bread or a chip on a sidewalk—a great deal of what it means to be a corvid takes shape. As these rooks diligently guarded their bounty they demonstrated both the capacity to *anticipate* theft by others and the ability to act preemptively to ward it off. This, too, is no small achievement; in fact, it may even be the case that these pilfering and antipilfering activities are in some sense *fundamental* to what it is to be a corvid.

Most crows, it seems, spend a solid amount of time each day stealing from others. If the beach-foraging northwestern crows (*Corvus caurinus*) of Washington State are anything to go by, they do so in targeted and purposeful ways. Studying groups of crows foraging together for clams, worms, crabs, and other tasty morsels in the intertidal zone, Renee Robinette Ha and James Ha found that while the odd crow did rely exclusively on food it hunted or collected itself, the majority of birds combined this activity with efforts to steal food from their neighbors (Robinette Ha et al. 2003). No crows relied solely on stealing. As Renee explained: "Crows cannot steal for a living because there are not enough opportunities to steal enough big food items" (Schwarz 2003). It was, however, an incredibly widespread practice among them. In fact, it is so widespread that the researchers think that it is actually theft and accompanying efforts to ward theft off, rather than fear of predators, that primarily motivates foraging crows in their "vigilance" behaviors (that is, visually scanning the environment around them, rather than keeping their heads down in search of food) (Robinette Ha and Ha 2001). The crows they studied also stole rather indiscriminately from relatives and strangers, although they did seem to tailor their thieving strategies. When stealing from a more closely related bird, a crow often quietly approached and took the food, whereas when a less closely related bird was the target, theft often involved a noisy, squawking approach and a subsequent pursuit until the fleeing bird dropped the food (Robinette Ha et al. 2003).

In this context, crows were primarily stealing from others opportunistically, as the food was procured. But, importantly, corvids don't limit themselves to this kind of theft. In addition, they have become highly skilled at raiding one another's "caches," that is, the little bits of food and other items that corvids hide away for later. Almost all corvids cache: from jays hiding the seasonal bounty of acorns and other seeds, and ravens hiding bits of meat in the snow for later, to New Caledonian crows caching the tools they use to extract grubs from logs and branches (Klump et al. 2015).[2] With all this stealing going on, it makes sense that corvids tend to try to cache things away from prying eyes, and if they are seen, they often return later to move the item to a safer location. These are complex cognitive and social operations. It seems that corvids are not only keeping track of their own caches but also of which other birds saw them cache which items where. In one set of experiments, common ravens (*Corvus corax*) demonstrated this knowledge by acting to ward off the approaches of birds they knew to be knowledgeable about the location of a particular nearby cache, while ignoring the movements of ignorant birds (Bugnyar 2010, Bugnyar and Heinrich 2005).[3] Interestingly, another experiment with scrub jays (*Aphelocoma coerulescens*) showed that only birds who had themselves previously stolen from others took these kinds of preventative actions, implying, as the researchers note, "that jays relate information about their previous experience as a pilferer to the possibility of future stealing by another bird, and modify their caching strategy accordingly" (Emery and Clayton 2001, 443). In other words, as Nicola Clayton, professor of comparative cognition at the University of Cambridge, put it in our conversation, "it takes a thief to know one."

Many of these studies of caching behavior are, more accurately, studies of *stealing*: of pilfering and antipilfering strategies. This topic has been of particular interest to biologists not because they have a strong interest in questions of corvid morality but because of what crows might here reveal about their ability to attribute mental states to others, referred to in biology as possessing a "theory of mind" (ToM). In acting in the ways that they do, crows seem to demonstrate an understanding of other crows as mindful beings, subjects with their own unique "perceptions, attentions, intentions, and beliefs" (Bugnyar 2007, 15).[4] Far from simply responding to where another bird is looking or going ("reading behavior"), recent studies strongly indicate that these birds are attributing mental states to others, as demonstrated in work in which ravens took preventative measures to stop pilfering by birds they could not see but knew *might* be watching them (Bugnyar, Reber, and Buckner 2016).

Two of the main laboratories engaged in studying these complex interactions between pilfering and antipilfering—that of Nicola Clayton and that of Thomas Bugnyar at the University of Vienna—have reached a similar conclusion: this behavior may be the key driver in the evolution of the remarkable intelligence of corvids (Dally, Clayton, and Emery 2006; Bugnyar and Kotrschal 2002). Central to this possibility is the development of spatial and observational memory, which allow birds not only to relocate their own caches but to watch where someone else has cached, remember the location, and return later. As Clayton explained to me: "Caches are under much greater risk of being stolen if others can watch and learn than they are if they can only be stolen either at the time [of caching] or opportunistically encountered. Observational memory for caches has probably driven the increasing cognitive complexity of both stealing strategies and cache-protection tactics, because an individual bird is both the protector of its own caches and a potential pilferer of others." In other words, as Bugnyar and Kotrschal (2002, 193) put it: "this competitive game for food may fuel an intraspecific evolutionary arms race for deceptive and cognitive abilities."

This fascinating hypothesis places hiding and, of course, stealing at the center of our stories about how it is that crows became who they are. If Clayton and Bugnyar are correct, then perhaps it is pilfering and its prevention that have, more than anything else, enabled their complex cognitive and social lives. Corvid wakefulness is, at least in part, a product of and a preparation for theft—it is stolen property. Stealing is at the core of who crows are. In fact, the more I learn about these activities, the more comfortable I am labeling them as "theft." While this term surely has a variety of meanings and associations within diverse cultural, not to mention biological, contexts, it seems to me that it is not right to assume that to apply it to the activities of nonhumans is necessarily an anthropomorphic projection. We are, at the very least, in the same neighborhood here. Crows do seem to have a sense of theft: they understand and negotiate its social intricacies, its hostilities and niceties, its conduct and its prevention. They steal knowingly, deliberately, sometimes even carefully—certainly from one another but perhaps also from others, including humans. In making this point, my aim is not to slip into the unhelpful forms of moralizing that often accompany discussions of theft. Rather, it is to learn to see and appreciate in new ways what is at stake, what is *made possible*, by stealing. Whole modes of life—fascinating, rich, intelligent ways of being—have been stolen into existence, brought into our world in no small way through this particular space and practice of being with others.

FIGURE 2.1. Hawaiian crow. *Source*: Photo copyright Jack Jeffrey.

Chapter Two

SPECTRAL CROWS

Conservation and the Work of Inheritance

THE BIG ISLAND, HAWAI'I

I stood in the forest listening for crows. Listening and hoping, even though I knew that it was foolish. I had been led to this forest precisely because there were no longer crows here, because there were no longer free-living crows anywhere in Hawai'i. I knew that the last sighting of a crow had been made over a decade earlier and that these birds were now extinct in the wild. But as I stood in the forest, I couldn't help but listen and hope. I had read descriptions of crows in Hawaiian forests by eighteenth- and nineteenth-century naturalists and ornithologists, writing when these birds were still relatively common. George Munro (1944, 70) saw them in 1891 and provided a passing reference to their graceful movements below the rain-forest canopy: birds "sail[ing] from tree to tree on motionless wings." Standing in a forest at 7,000 feet elevation—in the heart of the region where they once lived—I imagined for a moment that I could see their feathered forms moving through the trees. I imagined what it would be like for the now eerily quiet forest, missing this and so many other species of birds, once again to be enlivened by such a charismatic presence.

These spectral crows were haunting a forest that is itself in many ways dying. Like many of Hawai'i's forests, this one is in decline for a number of reasons, principally because of the presence of introduced ungulates, like pigs, that uproot and graze down any new vegetation. Where once there had been a lush understory beneath a tall canopy of trees, all that remained now were old trees with no new growth to replace them and no understory to hold the soil together when it rained. The biologists I was traveling with called this a "museum forest"; others have called it a forest of the "living dead" (Sodikoff 2013). Either way, it too was perched perilously at the edge between life and death.

The Hawaiian crow (*Corvus hawaiiensis*) is known locally by its Hawaiian name, ʻalalā. Although these birds look very much like the more abundant continental species of crows and ravens found widely around the world, behaviorally they are quite distinctive. ʻAlalā are forest and fruit specialists who have not taken to scavenging. Instead, they eat flowers and fruit, insects, and occasionally other birds' eggs. While they were once a common sight in the drier ʻōhiʻa forests in the south and west of the Big Island, by the late nineteenth century ʻalalā numbers had started to decline drastically. As forests were cleared for plantations and ranching, and many of those that remained were severely degraded, the ʻalalā did not take to a life beyond the forest and instead remained bound to these disappearing places. At the same time, they had to contend with new avian diseases and predators like cats and mongooses, as well as with the lethal persecution of farmers and orchardists who viewed them as a threat to their crops (Banko, Ball, and Banko 2002; Ball et al. 2016).

Eventually, in 2002, the last of the free-living ʻalalā died. Initially, only a handful of crows survived in captivity. As a result of years of captive breeding, however, there are now roughly 125 ʻalalā held in two facilities on the Big Island (Hawaiʻi) and Maui as part of a conservation collaboration between the state and federal governments and the San Diego Zoo. These captive facilities were the only home for the species for over a decade, until in 2017 eleven birds were released back into the forests of the Big Island. Then, in 2018, a further ten birds. Today, all twenty-one released birds are still surviving, and conservationists hope that from this tentative, fragile, free-living population, the ʻalalā might once again thrive on this island. However, as Jackie Gaudioso-Levita, the coordinator of the ʻAlalā Project, has noted, it will take years and years of releases and monitoring to reestablish the species: "the real milestone is decades into the future when the offspring of these birds are successfully breeding in the wild" (Hurley 2018).

This chapter explores the past and possible futures of this critically endangered corvid. It is based primarily on fieldwork conducted in 2011 and 2013, when the prospect of releasing ʻalalā back into the forest was still a somewhat distant hope. Or, more accurately, it was a hope for some people and a source of ongoing concern for others—giving rise to considerable animosity, in particular between conservationists and pig hunters. Much of this conflict centered on a particular place, the Kaʻū Forest Reserve in the south of the Big Island, which in 2013 had been identified as the most likely site for establishing a large, long-term release program for the ʻalalā. This release, however, would involve fencing and the removal of pigs (and so possibilities for pig hunting), so that a large section of the reserve might be revegetated. This debate drew in a broad cross-section of the local community, including many Kānaka Maoli (Native Hawaiians). Some Kānaka supported the conservation plan, others vocally opposed it, connecting it to the loss of hunting as a traditional cultural practice, the further undermining of their access to ʻāina (land, place of nourishment), and larger ongoing processes of colonization in the islands.

In exploring the story of the ʻalalā, this chapter focuses on how we inherit and inhabit the legacies of the past to shape possible futures. These inheritances take many forms, from genetic material and the broader landscapes and ecological communities that we are born into to the historical events, cultural traditions, and relationships that we retell, reenact, and remember. These inheritances haunt the present, for better and worse, in often unexpected ways. A key part of this haunting is the way in which the particular histories that we tell, that we inhabit, animate our understanding and action. Histories are not *of* the world but *in* the world, as Donna Haraway (2016, 14) reminds us of stories in general—they are part of the world's ongoing emergence. How we tell the past, as well as which pasts we tell, plays a powerful role in structuring what is nurtured into the future and what is allowed or required to slip away.

In a time of ongoing extinction and colonization, a time in many ways characterized by interwoven patterns of biological and cultural loss, what does it mean to inherit *responsibly*? My contention is that in a "postnatural world"—one that refuses the dangerous illusion of a wilderness that can be returned to, a nature "out there" divorced from human life—conservation must be rethought as a *work of inheritance*. This focus on the past may seem odd given the deeply uncertain future of this species at the current time, yet it is only out of these inheritances that our worlds can be crafted. It is only through paying attention to these processes that, as ku'ualoha ho'omanawanui (2014, 679) has noted, we are able to appreciate the many ways in which "the light of understanding shines from the past and illuminates our continuing work today and tomorrow." Inheritance is a worlding work.

When the 'alalā were finally released back into the forest in 2017, it was to a different part of the Big Island: the Pu'u Maka'ala Natural Area Reserve on the southeastern slope of Mauna Loa. This site had already had decades of management in collaboration with neighbors and partners, including fencing and ungulate removal. It is also a dedicated wildlife conservation area and so does not have the challenges associated with working with hunters and other land users. It is, however, relatively small. As such, the Ka'ū Forest Reserve is very much still a part of future plans for the 'alalā, and work on the fencing has begun in the intervening years. As Donna Ball from the USFWS put it to me, "the Ka'ū site is the big one, long-term . . . it really is an important habitat for the birds."[1] Despite these important changes in circumstances for both 'alalā and local people, because the events that I recount here took place in 2013, I have kept my focus primarily on this earlier period. In this particular story of contestation over the 'alalā we encounter valuable lessons for understanding and ultimately for crafting worlds out of the multiplicitous and often conflicting inheritances we carry with(in) us.

GHOSTS AND CO-BECOMING AT THE EDGE OF EXTINCTION

We don't know when it was, or where they came from, but at some point in the deep history of Hawai'i, crows appeared. As the islands in this volcanic chain rose above the sea, one by one countless plants, animals, and other species arrived by wave, wind, and wing and settled in. A diversity of life broke forth. Animals and plants adapted, coevolving with others over millions of years. Completely free of mammalian predators, for the longest time these were islands of immense avian diversity. Fossil records indicate that there was once a range of large, flightless birds in the islands (Steadman 2006). It is likely that in earlier times, many of these birds played important ecological roles as pollinators or seed dispersers for local plants.

Today, however, most of these birds are long gone. Some species were extinguished shortly after the arrival of Polynesian peoples in the islands, which may have taken place as early as 300 CE.[2] As people settled the islands, they cleared land for the farming of kalo (taro) and the other species that would make these places hospitable to Polynesian life, while also introducing rats that ate birds' eggs and nestlings and themselves hunting birds for consumption (perhaps especially larger, flightless species). Other avian species, like the ʻalalā, survived through this period only to be decimated by the waves of European, American, and other explorers, settlers, and then colonizers that began arriving in the islands in the late eighteenth century, setting in motion a drastic scaling up of all of these impacts (Boyer 2008). Vast areas of forest and other lands were cleared or otherwise destroyed in Hawaiʻi for plantations and ranching and then for urban development, military use, tourism, and more. At the same time, huge numbers of new species arrived, including diverse ungulates that consumed the forests, predators that ate birds and their eggs, and diseases like avian malaria and pox. These combined impacts have devastated Hawaiʻi's birds. Of the 113 avian species known to have lived exclusively on these islands just before human arrival, almost two-thirds are now extinct. Of the forty-two species that remain, thirty-one are federally listed under the U.S. Endangered Species Act (Leonard 2008). It is not hard to see why Hawaiʻi is regarded as one of the "extinction capitals" of the world.

As a result, the ʻalalā is now the largest fruit-eating forest bird remaining anywhere in the islands—albeit only in the form of captive birds and the tiny free-living population released into the Puʻu Makaʻala Natural Area Reserve. The loss of all these larger birds matters profoundly for the plant and tree species that relied on avian seed dispersers, especially those species with bigger fruits and seeds. Under the rainforest canopy, wide seed dispersal can be a vital component of species survival. As birds carry seeds away from their parent trees, they spread genetic diversity, reduce competition, and provide safer places for germination. Many Hawaiian plants can also grow epiphytically, safe from browsing ungulates, if their seeds are deposited up in the canopy by birds. Research conducted by Susan Moana Culliney and her colleagues (2012) suggests that the ʻalalā may be the last remaining seed disperser for at least three plants: hoʻawa, halapepe, and the loulu palm. But dispersal is not just about movement. In addition, it seems that some of these seeds germinate better—or, in the case of hoʻawa, germinate only—if the outer fruit has been removed, something that ʻalalā once routinely did.

A long and intimate history of coevolution lies within these embodied affinities that bind together avian and botanical lives. Crows are nourished, plants are propagated, and in the process both species are, at least in part, constituted: their physical and behavioral forms, their *ways of life*, emerging out of generation after generation of coevolutionary intra-action. ʻAlalā haunt the forest in another way here. Beyond my own active imagination, their spectral presence is *inscribed* in the forest landscape. Plants call out to ʻalalā, their fruiting and flowering bodies shaped by past attractions and associations that no longer really exist. These botanical "calls" are not simply metaphorical; they are fleshy, embodied, evolved, and continually reenacted semioses emerging in the interplay between a caller and a receiver.[3] With the recent release of ʻalalā back into the forest, it is possible that these calls might once again be heard, taken up, made whole.

As ʻalalā populations have declined over the past decades, the plants bound up in mutualistic relations with them have likely declined, too. Halapepe and loulu palms are themselves rare or endangered. In addition, Culliney (2011, 21) notes, with regard to hoʻawa, that most of the trees encountered today are older and that there is now a "general lack of seedlings or saplings in the wild." It is quite possible that these plants are now what biologists call "ecological anachronisms": species with traits that evolved in response to a relationship or an environmental condition no longer present (Barlow 2000, Janzen and Martin 1982). The extent to which the decline of ʻalalā and the loss of so many other large species of birds has contributed to a reduction in the populations of these plant species remains a topic for future study. It is clear, however, that the absence of seed dispersers can only make the future of these plants that much more precarious. Here, we see that coevolution can switch over into coextinction, cobecoming into entangled patterns of dying-with.

The disappearance of ʻalalā has also been felt by local people. For some Kānaka Maoli, ʻalalā are an important part of their cultural landscape. Like most Hawaiian plants and animals, ʻalalā animate moʻolelo, mele, and oli (stories, songs, and chants) (Osorio 2014). This bird is also an ʻaumakua, or ancestral deity, for some families, and the plants and forests that might disappear or change significantly without their seed dispersers are also culturally significant in various ways (Culliney 2011). Many other local people are also drawn into this experience of loss. During my time on the Big Island in 2011 and 2013, when the ʻalalā was still restricted to the aviary, I interviewed biologists, artists, ranchers, hunters, and others—some of whom were lucky enough to remember, and so now miss, the dramatic presence of these birds in the forest. Many of these people were trying in their own ways to reckon with the affective burden of living in a place in which crows are no longer present, a place in which (paraphrasing one biologist) we have lost the most intelligent and charismatic component of our forests.[4] Although, today, the chance of encountering one of the handful of birds now released is exceedingly slim, it is hard to overstate the transformative power of the knowledge of their presence—out there in the forest—for some local people.

Here, crows, plants, people, and others are tangled up and at stake in one another in a diverse range of ways. It is the particularly historical character of these entanglements that interests me here, though: the way in which living beings—ʻalalā, hoʻawa, diverse humans, and more—carry traces of the past that shape everyone's futures. We are all woven through with the past in this way: our own past, but also that of our forebears, whose relationships and achievements we inherit in our genes, our cultural practices, our languages, and much more. Life is, at a fundamental level, grounded in rich patterns of biocultural *inheritance.*

As Donna Haraway (2014) has noted—drawing on work in developmental systems biology—we are all "lichens": beings composed as, and out of, entanglements of diverse others, shaped by inheritances much more complex than a genetic blueprint. The "cultural" and the "biological," the "evolutionary" and the "developmental," cannot be neatly teased apart here. This is an understanding that has only really gained traction in the biological sciences over the past couple of decades. Such an understanding contrasts with that of the so-called new synthesis of the early twentieth century, in which inheritance tended to be understood through an almost exclusive focus on the transmission of *genetic* material between generations and was largely divorced, conceptually, from developmental processes. Today this understanding has been drastically unsettled by the realization that it is not just genes (along with epigenetic factors) that are inherited in meaningful and vital ways (Griffiths and Gray 2001).[5] A whole range of factors that might be thought about as "environmental" (from the very particular developmental space of the womb to the larger ecosystem) and factors that might be called "cultural" (behaviors, languages, and more) are in an important sense passed between generations, enabling the continuity of particular ways of life (Oyama, Griffiths, and Gray 2001; Jablonka and Lamb 2005). Some of these inheritances are linear—from biological parent to offspring—but they are also more than this: they are radically multivalent and radically multispecies.[6] Who we all are as individuals, as cultures, as species, is in large part a product of generations of co-becoming in which we are woven through with traces of all of our multispecies ancestors.

An appreciation of these kinds of entanglements makes it easier to understand why a species like ʻalalā cannot be neatly excised from our living world. Extinction always takes the form of an unraveling of co-formed and -forming ways of life, an unraveling that begins long before the death of the last individual and continues to ripple out long afterward: hosts of living beings—human and not—are drawn into extinctions as diverse heritages break down or are otherwise transformed.[7] There are no solid lines here between people, crows, and their environments, between evolutionary and cultural entanglements: relationships and affinities cut across simple divides, moving back and forth with ease. The traces that we leave behind in one another remind us that conventional Western notions of "the human" as a being set apart from the rest of the living world have always been illusory (Plumwood 1993). In Anna Tsing's (2011, 144) terms, "Human nature is an interspecies relationship." Or, as it is sometimes succinctly put by Kānaka Maoli, the people arrived as Polynesians, but the islands made them Hawaiian.

SPECTRAL CROWS AND THE PROMISE OF RETURN

As I traveled, observed, and talked with a range of people on the Big Island in early 2013, I encountered another important site in which the absence of crows was helping shape future possibilities for everyone. At the center of this story is the Kaʻū Forest Reserve—the forest in which I stood listening and hoping for crows. Early in my trip, I traveled high up into this area with a group of conservationists and state and federal land managers, a two-hour drive on a very bumpy dirt road that crossed old paddocks, forested areas, and cooled lava fields that stretched out black into the distance as far as the eye could see. Just a few months earlier, the state government had released its management plan for the area. At the core of the plan was a proposal to fence 20 percent of the reserve, roughly 12,000 acres (DLNR

FIGURE 2.2. Degraded forest, missing most of the understory plants. Ka'ū Forest Reserve, island of Hawai'i. *Source*: Photo by author.

FIGURE 2.3. Maps used to illustrate forest damage to local people. Ka'ū Forest Reserve, island of Hawai'i. *Source*: Photo by author.

2012). The fenced section would still allow human visitors, but all the pigs inside would be killed. As the forest recovered, it was anticipated that it would be a future release site for ʻalalā, while also contributing to the conservation of a range of other endangered species and ensuring that erosion was minimized so that the forest might remain a healthy water catchment for local communities.

Not everyone supported this plan. Although its drafting involved more than a year of serious community consultation, it was greeted with hostility by some. The most vocal opposition came from hunters—some of them Kānaka Maoli—who did not want to see a fence built and the pigs they hunted removed from the area. Of course, hunters are a diverse crowd in most places, and this is certainly true in Hawaiʻi. In this context, opposition to fencing was grounded in a range of understandings, values, and histories. On the surface, the most prominent opposition to this fence was justified by the notion that there is not enough accessible public hunting land in Hawaiʻi and that too much land is already "locked up" in conservation.[8] In short, for these people it is often simply a question of whether the interests of birds, plants, and the islands' many endangered snails and other invertebrates should take priority over those of humans. In addition, hunters often challenge the notion that pigs and other ungulates damage the forest, some even arguing that pigs actually play the positive ecological role of tilling the soil and rooting out weeds.[9]

The three conservationists who led our little expedition to the Kaʻū Forest Reserve that day were all locals, born and raised in the district of Kaʻū. Both John Replogle, a former ranch hand, longtime hunter, and conservation convert, and Shalan Crysdale, an ecologist, were working for the Nature Conservancy. Nohea Kaʻawa, a Hawaiian woman with deep family roots in the area and a degree in Hawaiian studies, was working as a community outreach and education officer for the state government. Together, they played a central role in the drafting of the new management plan for the area, especially the community-engagement process. As part of this process, they took numerous groups of locals, including many hunters, up to the section of forest that the state was proposing to fence. After visiting the site, many hunters who were initially skeptical agreed that fencing is a good idea, partly because the visit impressed upon them just how remote the area is and therefore inconvenient for hunting (especially as it cannot ordinarily be accessed by vehicle) but also because they were able to see the extent of the damage that ungulates were doing to the forest.

During these site visits, John, Shalan, and Nohea also spent a lot of time talking to local people on the long drive up and back. John explained to me that one of the ways in which he conveyed the significance of the extinction of the 'alalā to local people was to draw a direct comparison between the loss of this species, on the one hand, and the potential loss of the language and culture of the Hawaiian people, on the other. In making this contrast, John was tapping into a topic of ongoing importance in the islands. While 'Ōlelo Hawai'i (Hawaiian language) and diverse cultural practices—from hula to voyaging—are today in the midst of what is often referred to as a renaissance, for most of roughly the past 125 years since the islands were colonized by the United States, these integral facets of Hawaiian life have been prohibited and/or actively denigrated.[10] In referencing this ongoing history John aimed to highlight the importance of inherited diversities of all kinds, of sustaining them into the future, allowing people to connect with the loss of a bird that, for some, had come to seem insignificant.[11]

Of course, some hunters opted not to go on these site visits, and others remained unconvinced. Many of these people continued to oppose the fencing and removal of pigs from this area. Some of the most vocal opponents were a small group of Kānaka Maoli hunters. For many Native Hawaiians, pig hunting is understood as a core traditional practice that ought to be widely supported as part of the continuity of Hawaiian culture. In conversations with these hunters, as well as in online discussion forums, I encountered repeated references to this point of view. For them, any effort to remove pigs and limit hunting was seen as a violation of their traditional and customary rights, protected by the Hawaiian constitution (sec. 7).[12]

In recent years, however, the notion that pig hunting is a traditional cultural practice has been thoroughly problematized. Detailed historical studies by the Hawaiian cultural experts Kepa and Onaona Maly indicate that before European arrival, pigs were kept close to home, and they were also distinctly different animals: the smaller Polynesian variety, not the large European boars now found widely throughout the islands. The primary hunting that likely took place at that time was bird hunting, mainly for feathers used in royal ornaments and clothing (Maly and Maly 2004, 152).[13] With this information fresh in my mind, I expected conservationists to readily dismiss claims by hunters to "tradition," but I found that this was not the case. Instead, almost all the conservationists I met with noted that this shorter history did not invalidate claims to continued hunting. Many noted that the length of time required to make something "traditional" was uncertain, that culture is not static, and that several generations of hunting is certainly long enough to establish family traditions—forms of identity and culture—that ought to be respected wherever possible. In short, they recognized in their own way that, as James Clifford (1986, 10) has famously put it: "'Cultures' do not hold still for their portraits."

But something else was happening here, too. Several of the conservationists that I spoke with quickly mentioned this historical research when the topic of pig hunting came up. Although they were clear that this did not mean that hunters had no claim to continue hunting, it certainly changed the *nature* of that claim. In noting that the pigs are different from those originally brought to the islands by Polynesians and that the practice is more recent than sometimes thought, a break with the past is effected in which fencing and pig removal are conceptually separated from contentious questions of Kānaka Maoli customary practices and rights. Different histories create different continuities and ruptures, with all their attendant political and ethical consequences (Bastian 2013). Importantly, however, it was not just *haole* conservationists making this claim.[14] In fact, some of the people who made it most strongly to me in interviews were Kānaka Maoli who see the removal of pigs from at least some areas of forest as essential to the conservation not only of the environment but of a rich notion of Hawaiian culture, too.

The desire of some conservationists to conceptually separate pig hunting from traditional Hawaiian culture was an effort to *depoliticize* plans to remove pigs. This effort was, in large part, a response to the prominent role that the occupation and colonization of Hawai'i by the United States was playing in some of the most vocal opposition to fencing in Ka'ū. With the occupation firmly in mind, for some hunters the proposed fence was positioned as one more "land grab" in a long history of taking.

The last monarch of the sovereign nation of Hawai'i, Queen Lili'uokalani, was overthrown in 1893 by a group of wealthy settlers with the aid and support of members of the U.S. government and its military. Through a complex series of events over the next five years, Hawai'i became a territory of the United States and fifty years later was made a state. Although there was some attempt, both in the lead up to the overthrow and afterward, to provide maka'āinana (commoners) with some form of property rights in small parcels of land, this never really worked out in their favor (Banner 2007, Silva 2004): from the Great Mahele of 1848, through subsequent decades of dispossession and annexation, until, in J. Kehaulani Kauanui's (2008, 75) words, "Hawaiians and their descendants [had become] largely a landless people."[15]

We are reminded here that at the heart of colonization is a profound and ongoing process of fracturing inheritances. As Jonathan Kay Kamakawiwo'ole Osorio (2002, 3) has noted: colonialism in Hawai'i worked "through a slow, insinuating invasion of people, ideas, and institutions." These processes undermined and prohibited the use of 'Ōlelo Hawai'i, as well as hula and diverse other expressions and practices of Hawaiian culture, producing a significant "intergenerational rupture" (Tengan 2008, 67). While, as Osorio and other scholars have documented (Silva 2004), many Kānaka resisted, fighting "this invasion with perplexity and courage," ultimately "that colonialism literally and figuratively dismembered the lāhui (the people) from their traditions, their lands, and ultimately their government" (Osorio 2002, 3).[16]

For people inhabiting this history, endangered species conservation (especially of the state-sanctioned variety) can often be regarded as one more excuse to take away people's rights to access or use land. As one hunter put it, environmentalists are "always using something endangered to the i[s]lands for try grabb land."[17] Importantly, these people often do not trust the intentions of government agencies in this area, viewing any fencing as the beginning of a slippery slope toward complete loss of access. As another hunter put it: "Environmentalist want to eventually take it all away and fence it in! They're starting with these areas, and will start working on more. The alala, water shed, native plants, etc. is just a smoke screen to grab more land!"[18] There is something very familiar about these views. In many parts of the world—including the continental United States—hunters express similar concerns about conservation (Emery and Pierce 2005, McCarthy 2002). But there is also something distinctly Hawaiian about them; there are clear echoes of the Great Mahele and acts of subsequent dispossession here, as well as frequent references or allusions to traditional rights. Perhaps most importantly, however, these arguments by hunters often explicitly challenge the authority of the state government and, certainly, that of the federal government—illegal governments from this perspective—to exercise any authority in the management of these lands and resources. In this way, the conservation of the ʻalalā is connected to a much broader—deeper and ongoing—process of colonization that has profoundly fractured not only access to land and the right to hunt but a host of other Kānaka Maoli inheritances.

This connection between conservation and colonization has significant consequences for ʻalalā and other endangered species. Once a proposal like the Kaʻū Forest Reserve Management Plan has been framed in this way, those who speak in its favor are frequently positioned as endorsing colonization. As Shalan put it to me in an interview: "To be for the plan is to be for the overthrow."[19] In this context, publicly supporting conservation—as a Kanaka Maoli or anyone else—requires one to enter into what another local called the "raging fire of emotion" that surrounds the occupation and subsequent colonization of the Hawaiian Islands. In this light, ʻalalā themselves become an enemy of at least some of the Hawaiian people. What's more, the birds' movements through the forest become suspect as hunters and landholders fear that each time ʻalalā move beyond the fenced area (especially if they are nesting), the fence will expand with them. And so, the ʻalalā is imagined as a Trojan horse whose conservation facilitates further loss of land and rights. For some, they are not a biological inheritance to be valued and cherished but a symbol and a powerful material enabler of a broader colonizing process, a broader fracturing of Kānaka inheritances. It should come as no surprise that in this climate, conservationists had real fears that any birds released into the Kaʻū Forest Reserve might be targeted by some hunters.

INHERITING THE WORLD

Toward the end of my trip to the Big Island in 2013, I met with Hannah Kihalani Springer, a kupuna (elder) who lives in the district of North Kona. Hannah is deeply knowledgeable about Hawaiian history and culture and about hunting and conservation, so I was eager to hear her thoughts on the past and future of the islands. Sitting in her living room in her family's homestead, we talked about conservation, politics, sovereignty, ranching, and, of course, ʻalalā. Hannah is lucky enough to have seen free-living ʻalalā, sometimes in large gangs, throughout her early life. She recalled: "When we went into certain sections of our lands, it was the norm to see crows. From the fifties all the way through to the mid-seventies. . . . It was January 1, 1977—and I only know the date because my mother's birthday was January 1, and I had gone to pick maile [a highly fragrant plant] for her birthday—when two adults and a young bird came and worried me. That was the last close-up encounter that I had with wild crow." Hannah is a passionate and active conservationist, a past president of the Conservation Council for Hawaiʻi. Like many other people with whom I spoke, she felt that in some places pigs and other ungulates need to be fenced out and removed for conservation. But she also believes that room has to be made for hunters—her family hunts, and in the past she hunted too. And so, like others I spoke with, she felt that the government could do more to facilitate access to existing state land for hunting.

In contrast to those Kānaka Maoli who strongly emphasize the place of pig hunting in their culture, Hannah noted that the islands' forests are alive with a diversity of plants and animals, all of which have their places in Hawaiian stories and culture. In this context, she argued that a singular focus on pigs is not helpful. In her words: we need "the larger context that is much more diverse and dynamic. . . . When we so diminish the conversation we're diminishing the Hawaiian experience and the Hawaiian culture. The forest is important for the myriad characteristics that comprise the whole." Other Kānaka Maoli I spoke with who shared this view referenced another history in their arguments about the need to hold onto a diversity of plants and animals in the forest. These people referenced the *Kumulipo*, a creation chant that connects the Hawaiian people, through a shared kinship, to the gods and the generation of all life (McDougall 2016, Nākoa Oliveira 2014).[20] For these people, removing pigs from portions of the forest to aid in the conservation of ʻalalā, other endangered birds and plants, and the watershed was seen as essential for the protection of Hawaiian life and culture. In this context, it made sense to many of them that in a place like the Kaʻū Forest Reserve, 20 percent of the land might reasonably be set aside from the damages of pigs.

The diversity of views within the Kānaka Maoli community matters. As is the case around the world, conservationists' narratives in Hawaiʻi often selectively highlight those portions of the Indigenous community whose views align with their own projects (Sodikoff 2013, 155). Or, as Jonathan Goldberg-Hiller and Noenoe K. Silva (2011, 430) have noted, citing one example of relevant Hawaiian state legislation, these narratives often leverage a broad generalization in which "ecological balance is seen as coterminous with the "great cultural, historical, and spiritual significance [of a given species or place] for many native Hawaiians, native Hawaiian practitioners, and others who value the Hawaiian culture." On the other hand, however, the voices of Kānaka Maoli who support various government-sanctioned projects—from conservation to geothermal power and the construction of new telescopes—can also be drowned out by those who oppose them.[21] All of these issues are complex; all of them are grounded in diverse possibilities for inheriting legacies of the past and crafting more livable futures (Goodyear-Ka'ōpua 2017, Iwashita 2016).

Speaking with Hannah that day, I was reminded again and again that the histories that we tell are themselves *acts* of inheritance. Which is to say, that the aspects of the world that we nurture into the future are, in more or less significant ways, shaped by how we understand and tell the past. Histories structure our understandings of what particular continuities mean and why they matter. All of the rich cultural and biological inheritances that constitute our world are at stake, to a greater or lesser extent, in the histories that we weave out of, and into, this forested landscape.

There is an important dynamic at work in inheritance here that deserves further attention. As Jacques Derrida (2004) has noted, at its simplest level inheritance seems to be about continuity and retention: taking up the past and carrying it forward into the future. Of course, much of this inheritance is not actively chosen: we are thrown into our heritage; in Derrida's terms, it "violently elects us." But this is not the end of the story. Derrida reminds us that in any act of inheritance there is also transformation. He is primarily interested in what it means to inherit traditions, languages, and cultures, and he notes that while all continue from generation to generation, they are living heritages not fixed once and for all. It is this "double injunction" at the heart of inheritance that Derrida emphasizes, describing the act of inheritance as one of "reaffirmation, which both continues and interrupts" (Derrida, Roudinesco, and Fort 2004, 4).

Cristina Bacchilega (2007, 2) offers a related sense of inheritance in her discussion of narrative traditions in Hawai'i, noting that a tradition is "an *ongoing* process rather than a naturalized inheritance, and one that hinges on the complex negotiations characterizing its individual recollections and performances." Here, the performer must take up responsibilities to the past but also to the present and the future. In this context she argues: "It follows that the opposite of tradition 'is not change but oppression' ([Glassie] 396). Historical violence, in other words, is at the core of the rupture of tradition." This is a process that is all too familiar to those in the Hawaiian Islands, where Kānaka Maoli inheritances have been undermined, denigrated, and denied, while these connections to the past are simultaneously utilized in an effort to position Hawaiian people as "relics of the past," out of touch with the complexity of the contemporary world and its issues. Critical Kānaka scholarship by Noelani Goodyear-Kaʻōpua and others insists on a different kind of relationship between pasts and futures, one that refuses monological notions of progress and their singular, exclusionary, "settler futurities" (Tuck and Gaztambide-Fernández 2013). Instead, the past is seen as a vital and vibrant heritage: "our genealogies are a backbone stretching to the very inception of these islands," one that might ground "a wider possibility of movement, a more supple way to navigate through the world" (Kuwada 2015). And one that might, in turn, open up space for diverse "Indigenous futurities" that are, as Goodyear-Kaʻōpua has noted, not so much (or at least not only) "obstructions on a march to '*the* future'" but efforts to protect "the possibilities of multiple futures" (2017, 186).

The dynamic of retention and transformation at the heart of inheritance—a dynamic that carries the past into the future(s)—extends well beyond these diverse spaces of human languages and cultures. All living beings are involved in their own forms of life- and world-shaping inheritance. Evolution by natural selection—that great engine of new ways of life—is grounded in forms of inheritance that simultaneously retain the achievements of the past while constantly transforming them to produce new variability. This variability arises through recombination, mutation, and other forms of transformation and is the stuff of future change and adaptation. Moving beyond the narrow genetic reductionism commonly found in neo-Darwinian accounts of evolution, we are reminded that these are lively and varied processes in which diverse heritages move between organisms in a range of different ways to shape bodies and worlds.

In this context, the fundamental structure of life is one of inheritance. Darwin knew something like this when he drew a comparison between language and biological species, with an emphasis on the way in which both are at their core *genealogical*: seemingly "individual" languages and "individual" species are in reality simply moments within longer historical lineages (Grosz 2004). Here, life takes shape through the constant generation of variability, only some of which "sticks," only some of which is retained and so incorporated into the larger collective (be it a language, a species, or indeed a culture). As Derrida succinctly put it: "Life—being alive—is perhaps defined at bottom by this tension internal to a heritage, by this reinterpretation of what is given" (Derrida, Roudinesco, and Fort 2004, 3–4).[22]

Inheritance is a productive concept for extinction studies and the broader environmental humanities. It is a concept with a long and rich history across the biological sciences and the humanities more broadly, including in diverse spaces of Indigenous scholarship and activism. Reading these literatures alongside and through one another, we are able to begin to develop an appreciation for entangled *biocultural* inheritances in which the movements of genes, ideas, practices, and words between and among generations are all tangled up with one another, unable to be isolated into separate processes or channels of inheritance. If we scratch the surface just a little, these entanglements are palpable in Hawai'i's shrinking forests: as the island's biotic diversity continues its long role in helping nourish and shape local cultures, cultures that are, in turn, remaking those ecologies and the futures of their many inhabitants.

Thinking in this entangled way draws us, inexorably, into an understanding of the ethical work of inheritance. While it may be a gift or a burden, a heritage is always a responsibility. Something to be dwelled with—to be honored and acknowledged, even if not always avowed—placing us in relationship, perhaps under obligation, to those who have come before as well as those who will follow after. Where species, ecologies, and cultures are in processes of ongoing and dynamic change, much of what is and is not passed on is not up to any of us. Where we can and do play a role, however, the question is usually the same. Never simple, never clean: *What* is to be lost and what retained? Which losses will we accept, and in the name of which continuities (and vice versa)? From within a time in which so much biocultural diversity is being threatened and lost, often violently, what does it mean to inherit responsibly, and how might we live up to our inheritances?[23]

One of the many things I learned from Hannah that afternoon was that responsible inheritance is necessarily grounded in a recognition of and an attentiveness to multiple voices, with their diverse histories and imagined futures.[24] As our conversation was coming to a close, Hannah and I drifted into a discussion of the sovereignty movement in the islands. She told me about a relative of hers, deeply committed to Hawaiian sovereignty, who worked for the state government as a biologist. When asked about the incompatibility between her politics and her employment, this relative would say that she was conserving Hawai'i's biotic diversity so that when and if sovereignty comes, the people and the land are in the best possible condition for it. Although Hannah didn't explicitly state it, it seemed to me that she too shared this general view. She went on to say:

> The conclusion that I've arrived at is: 'I am a citizen of the land.' We have lived on this land, as I've described to you, since before Cook's arrival. And, we've seen chiefs rise and fall, we've seen an island nation born and die before its time, elected and appointed officials come and go, but here we stand. I'm less interested in the constitution that binds us or the flag that flies over the land than I am in the quality of life on the land. So, if there are elements within whoever's constitution it is that allow us to preserve and pursue the righteous management of the resources that we call home, then I am happy to pursue those. . . . I am loyal to this land. Whatever flag flies over it is one that I am willing to use the resources of to continue to be a citizen of this land.

Hannah's position is one of hope, within which resides a profound responsibility to both the past and the future. Hannah has not forgotten the events of 1893. But she wants to inherit this history in a way that refuses to regard support for conservation as necessarily support for an illegal occupation. She wants to inhabit the history of these islands, her and her family's history, in a way that holds open possibilities for the landscape and the people who are a part of it into the distant future. In short, she is proposing that we might care for 'alalā, *and* for Hawaiian culture and sovereignty, *and* for the rest of the land and its people.[25]

Of course, there will always be compromises and challenges here, and they will likely always be unequally distributed. But I am inspired by Hannah's effort not to abandon any of these inheritances, to pay attention to their entanglements, and to take on the work of nourishing them as a responsibility to the past and the possible futures to come. Here, I think we see that responsible inheritance requires that we engage with others—their histories, their relationships—to hold open a future that does not forget the past or attempt to reconstruct it but rather inherits it as a dynamic and changing obligation that must be lived up to for the good of all those who do or might inhabit it.

This is what Deborah Bird Rose (2004, 24) has called "recuperative work," work that begins from the conviction that:

> There is no former time/space of wholeness to which we might return or which we might resurrect for ourselves. . . . Nor is there a posited future wholeness which may yet save us. Rather, the work of recuperation seeks glimpses of illumination, and aims toward engagement and disclosure. The method works as an alternative both to methods of closure or suspicion and to methods of proposed salvation.

In this context, "taking care" is always a historical and a relational proposition; if we're doing it right, care always thrusts us into an encounter with ghosts, our own *and* others'. Some of us live in worlds haunted by evolutionary ghosts: anachronistic plants and lost seed dispersers. Others live in worlds haunted by the wrongs of 1893 and dreams of a sovereignty to come. Others remember ʻalalā in the forest when they were children or are tied to *this* bit of forest by memories of a grandfather who taught them to hunt. Responsibility resides in a genuine openness to these diverse voices with all their complex pasts and futures. In fact, it is precisely because these pasts and futures are so inseparable from these voices that what is required here is actually openness to a multiplicity of *worlds*.[26]

But, importantly, care and responsibility necessarily draw us out beyond the arbitrary and unworkable limits of a purely human space of inheritance. In short, "ours" aren't the only hauntings that constitute worlds. Some plants live and are now disappearing in worlds haunted by ʻalalā; some crows are drawn, *called*, to a forest beyond the aviary. Paying attention to a multiplicity of worlds means recognizing that nonhumans are not simply resources to be conserved or abandoned, inherited or cast aside, on the

FIGURE 2.4. Captive 'alalā (*Corvus hawaiiensis*) on the island of Hawai'i. *Source*: Photo by author.

basis of whether current generations of humans happen to want them around. Rather, ʻalalā, hoʻawa, and others are themselves constituted through immense processes of intergenerational life, the cumulative achievement of multispecies entanglements, adaptation, and inheritance across vast periods of time. As such, their own ongoing dramas, as well as those of the many other forms of life that have already made and might yet still make worlds with them, demand our respect and gratitude.

In paying attention to some of the diverse ways that nonhumans inherit their worlds, we become aware of just how much is at stake in extinction. For example, it is thought that in captivity the once remarkable vocal repertoire of ʻalalā—their raucous calls and mournful songs—may have been diminished. Perhaps this is because they have had less to talk about, or perhaps juvenile birds simply haven't been exposed to enough chatter from their elders. Similarly, knowhow about predators and how to avoid them, as well as the local environment and its food sources, may not have been passed between generations in captivity, potentially affecting the future survival of released birds. Conservationists are working hard to reinstill these understandings where they can, but in some cases it simply will not be possible (van Dooren 2016). In these and other ways, the long-accumulated heritages of the species—not just their genetics but learned behaviors that took advantage of generations of refinement and adaptation—have perhaps been undermined, to the detriment of any future life for ʻalalā in the forest. Here, too, we see that "historical violence . . . is at the core of the rupture of tradition" (Bacchilega 2007, 2). We see in the most tragic of ways that as a species, and as individual birds, ʻalalā are beings with their own inheritances. Much is at stake *for* them, not just *in* them at the edge of extinction. Furthermore, as we are seeing, the histories that humans tell play a significant role in shaping whether and in what ways ʻalalā are able to take up these heritages to contribute to the crafting of flourishing worlds for themselves and others.

Ours is a time of mass extinction, a time of ongoing colonization of diverse human and nonhuman lives. But it is also a time that holds the promise of many fragile forms of decolonization and hopes for a lasting multispecies justice. Here, the work of holding open the future and responsibly inheriting the past requires new forms of attentiveness to *biocultural* diversities and their many ghosts. But beyond simply listening, it also requires that we take on the fraught work—never finished, never innocent—of weaving new stories out of this multiplicity. Stories within stories that bring together the diversity necessary to inhabit responsibly the rich patterns of interwoven inheritance out of which we must together craft our shared worlds.

COOPERATING

Crows cooperate. Or rather, some of them do, some of the time. They might cooperate in the rearing of young, in mobbing a potential predator, in sharing a bountiful food resource, or perhaps in gaining access to a difficult one.[1] In experimental interactions with biologists, food is often the motivator that is used to explore the when, how, and with whom of corvid cooperation. On my first visit to the Haidlhof Research Station in Bad Vöslau, just outside Vienna, Austria, I saw one of the archetypical experimental apparatuses for studying cooperation in action . . . almost. The "loose-string" experiment involves a platform holding a tasty morsel of one sort or another that is out of reach of the test subjects but that might be pulled into reach by means of a string. The catch is that the string is not attached to the platform. Instead, it is looped through two hooks or pulleys. Both ends of the string need to be pulled simultaneously for the platform to move. If the string is only pulled from one end, the apparatus will simply unthread, leaving the platform out of reach. On the day of my visit, no loose-string experiment was actually taking place, but as Thomas Bugnyar and I walked around the outside of the aviary, past the platform setup, one of the ravens moved into position by the string. Thomas reached

---

FIGURE V.3.1. Illustration by Kirsty Yeomans (www.crowartist.co.uk).

FIGURE V.3.2. A common raven (*Corvus corax*) exploring a shoe at the Haidlhof Research Station in Bad Vöslau, Austria. *Source*: Photo by author.

down and put a piece of food onto the platform. The raven, watching closely, took hold of the string and began to pull. In the absence of a partner, however, it was to no avail.

This anticlimactic demonstration was my first encounter in person with the loose-string experiment, although I had frequently read about it. In formal studies using this apparatus, corvids have shown themselves to be capable of cooperating. But again, only sort of. Rooks (*Corvus frugilegus*) and common ravens (*C. corax*) are among the very few birds—indeed, very few species of any kind—who have been tested with the loose-string paradigm. It was at this research station in Austria that these experiments with the ravens took place (Massen, Ritter, and Bugnyar 2015; Seed, Clayton, and Emery 2008). Both species successfully operated the apparatus, again and again, with various combinations of individuals, to gain access to their edible rewards. Some pairs of birds did so on their first try, others required a little longer, but all seemed to get the hang of it relatively quickly.

But, the researchers wondered, what does this successful action really mean? How much do these birds *understand* about how and why this setup works? Do they *know* that they require a partner? Are they intentionally cooperating or merely achieving a cooperative effect "through acting apart together, most probably motivated by a mutual attraction to the apparatus and the food"? (Massen et al. 2015, 2).

Two important control tests are frequently used as part of the loose-string paradigm to explore precisely these questions. In the first of these tests—the "delay" test—one bird is given access to the apparatus while their partner is unable to reach the string. The bird who can access the string must delay their pulling to wait for their partner to arrive. In the second test—the "choice" or "solitary" test—in some cases the setup is slightly modified so that a single bird is offered an alternative apparatus as well, in which it can pull both ends of the string at the same time, or the setup remains the same and researchers simply observe whether a solitary bird bothers to pull on one end of the string without a partner present. In either case, these tests aim to determine if birds understand the importance of both ends of the string being pulled simultaneously for the platform to move.

Both the rooks and the ravens failed both of these control tests. The experiment with the rooks was conducted at the University of Cambridge by a team including Nicola Clayton. In this case, all birds pulled the string before their partner arrived, most of the time (delay test), and four of the six birds showed no preference for the setup that did not require a partner (choice test). However, the two remaining rooks—named Cook and Selvino—showed signs of having slowly figured out the trick to the choice test. In their first ten attempts, both rooks showed no clear preference for the platform that would yield results without a partner, but in the second batch of ten attempts, Selvino went for the "correct" platform every time, while Cook did so eight out of ten times (Seed et al. 2008, 1427). As with all animals, including humans, exposed to novel, cognitively challenging situations, some rooks fare better than others. The researchers concluded, however, that as a group the birds' behavior doesn't suggest that they understand the significance of working with a partner on this particular task.

Despite this result, the scientists at the Haidlhof Research Station who were working with common ravens initially suspected that their birds would fare much better. Although rooks and ravens are closely related—and likely possess many of the same cognitive capacities—their social systems are quite different. While both species form long-term breeding pairs as their primary social units, rooks tend to do so early on in life, whereas ravens require a territory to breed and so can spend as much as a third of their lives waiting for one to become available. During this time, they form large groups for "foraging, roosting and socializing," as well as groups within these groups, requiring that individual birds maintain "multiple highly differentiated affiliative and agonistic relationships with others" (Massen et al. 2015, 2). As a result, raven social relationships are thought to be quite similar to those of chimpanzees, possessing precisely the kind of "variable social network made up of competitive and cooperative relationships" that the researchers working with the rooks thought might be part of the explanation for their inability to master this task (Seed et al. 2008, 1421). But the ravens, too, failed both of these control tests. Specifically, they failed around 98 percent of the delay tests and 84 percent of the solitary tests. The researchers summed up the result succinctly: "These data suggest that the ravens did not understand the need for a partner to solve the task in this experimental set-up" (Massen et al. 2015, 6).

Why begin a discussion of corvid cooperation with an example that indicates, at best, a partial success? Fascinatingly, while the rooks and ravens seem not to have demonstrated a "robust understanding" of the significance of working with a partner on this task, this does not mean that the presence and identity of their partner was irrelevant to them. Far from it. In fact, both species performed far better in the initial (precontrol) tests when they were working with another bird with whom they had a friendly ("tolerant") relationship. What's more, the friendliness of those relationships was itself something that was dynamic, at stake, in these string-pulling experiments. In the case of the ravens, the researchers found that if the rewards from a successful string pull were not evenly distributed—that is, if one raven displaced the other and took both bits of food—the cheated raven was much less likely to want to participate in future iterations of the experiment if working with the same cheating partner. In some cases, cheated birds were generally more reluctant to participate in the exercise at all, but here too the researchers found that the specific partner mattered: "those that are paired with a friend resume cooperating earlier than those that are paired with a non-friend" (Massen et al. 2015, 7).[2]

There are no simple, general outcomes here. Different species, different individuals, and different histories, activities, and contexts give rise to very different possibilities. Much depends on the specific "apparatus," whether we're talking about a highly structured experimental protocol or free-living birds working together on a lonely country road—some as scouts, others as pullers—in the effort to move a dead animal to the safety of the verge.[3] The how, why, and with whom of cooperation matters. Indeed, as the raven researchers noted in their discussion: "It is possible that the apparent lack of understanding in the current study is specific for the paradigm used" (Massen et al. 2015, 7): the ravens may simply not have understood what they were being asked to make sense of and do. "Future research should reveal whether ravens in general do not understand the need for a partner while cooperating, or whether it depends on the paradigm used" (Massen et al. 2015, 7). Or were these birds simply not able to suppress their learned string-pulling behavior, even though they knew on some level that it would not yield results? "It is well documented that many animals find it very difficult to inhibit a learned response for even short periods of time in order to get food" (Seed et al. 2008, 1427).

Nonetheless, there is something curious going on here. Both rooks and ravens seem to be highly tuned in to whom they are working with, or at least alongside, but perhaps don't really understand the role of that partner (at least within this particular setup). In short, "even though they attend to who is sitting next to them, they do not seem to attend to what this individual is doing" (Massen et al. 2015, 6). We see a similar attention to the *who* of cooperation in other contexts where corvid social relationships have been studied: for example, when food is shared, grooming provided, or consolation offered after a conflict (Fraser and Bugnyar 2012; Szipl et al. 2015; Asakawa-Haas et al. 2016; Seed, Clayton, and Emery 2007). In some cases these seem to be tit-for-tat reciprocal interactions; in other cases they are long-term relationships of more general mutual cooperation and support (something certainly the case with breeding pairs but perhaps more broadly too). In this way, while these string experiments might point to a "failure" or inability of one kind, they also offer a glimpse into an equally interesting space of complex social behavior, of birds paying attention to one another—who cheats and who doesn't, who helps and who doesn't, who shares and who doesn't—and remembering and acting on this information in the future. Here too, perhaps we view another fundamental part of what it is to be a crow: to be a being profoundly attuned to and shaped by thick social relationships.

In the past couple of years, another study out of Bugnyar's laboratory at the Haidlhof Research Station has added new depth to our understanding of the way in which corvids attend to one another. Crows, it seems, have a strong aversion to inequity (Wascher and Bugnyar 2013). The study in question focused on both common ravens and carrion crows (*C. corone corone*). Here, birds who had previously been trained to exchange a token with researchers for a food item were given the opportunity to do so in the presence of another bird, who was also making exchanges and receiving rewards. The researcher holds the token in one palm while showing the reward to the bird in the other: a grape, a piece of cheese, or nothing at all. Cheese is a much more highly valued food item than grapes for these birds, as they are regularly fed the latter as part of their diet. The offer of nothing at all is, obviously, the least desirable possibility. These birds were required to take the token into their beak, hold it for at least two seconds, and then place it back in the researcher's hand. If they did all this, they would receive the reward on offer. The core of this experiment, however, is its social context. In some cases both birds were treated equally, while in other cases one was given favorable treatment: while one bird was exchanging for grapes, another might be given cheese or nothing at all for conducting the same exchange process. Likewise, while one bird was exchanging tokens for food, another might simply be given the same food as a "gift," without being required to do anything at all. Across each of these experimental contexts, the crows and ravens demonstrated their awareness of, and their dissatisfaction with, situations in which another bird was getting a better deal. They did so by refusing to take the reward when it was not fair or by reducing their participation in the whole process, refusing to take the token in the first place.

Interpreting these kinds of behaviors is a difficult task. It seems, however, that these crows and ravens possess a general sense of fairness, an expectation that they receive equal treatment (something that they seem to share with some primates, dogs, and a few other species that have been tested in similar ways). It isn't immediately clear *why* they should hold such an expectation, though: corvid social groups certainly aren't spaces of free and equal treatment. There are thought to be definite dominance hierarchies, "pecking orders," within any stable group of corvids.[4] Perhaps the "social relationships" being negotiated in and through this experimental apparatus include the researcher in a central way? Perhaps equity of treatment, of access to food and rewards, is an expectation for these crows specifically in their dealings with humans, or even with these particular humans (whom they would be able to recognize individually and interact with regularly)? In other words, perhaps this is not an expectation that extends to (m)any of their crow-crow relationships?[5]

If this is the case, then while equality certainly does seem to be a thought that crows are capable of thinking—more than this, of enacting—it might also be an emergent product of the broader forms of human-crow sociality that take place in the particular context of shared lives inside research environments (or at least a possibility that has been refined and developed in new ways there). In these environments, clever birds must learn to live inside particular regimes of experimentation and reward, learn to inhabit other kinds of social relationships. Again, the specificity of the apparatus matters. It profoundly shapes the social possibilities, and ultimately the worlds, that might be imagined and enacted. There is nothing "unnatural" at work here. As Vinciane Despret (2008b) has argued so eloquently, no apparatus reveals the "true" animal; rather, each renders an animal capable, and so possible, in new ways. Crows, it seems, are capable of recognizing and insisting on something like equality in their cooperative interactions—at least some of the time.

FIGURE 3.1. A house crow (*Corvus splendens*) in Hoek van Holland, the Netherlands. *Source*: Photo courtesy of Ben Dielissen.

Chapter Three

UNWELCOME CROWS

Hospitality in the Anthropocene

ROTTERDAM, THE NETHERLANDS

On the morning of my second day in Hoek van Holland I went in search of crows. Corvids aren't hard to find in this small town on the west coast of the Netherlands. Jackdaws (*Corvus monedula*), in reasonable numbers, are a permanent fixture in the main town square during the daylight hours, eating crumbs and other discarded food. But I was in search of considerably rarer corvids. I was looking for house crows (*Corvus splendens*). At the time of my visit in early autumn 2014 there were thought to only be seven or eight individuals of this species left in town and, by extension, in the country or indeed anywhere in Europe.

Living, as we are, in a period of such incredible biodiversity loss, one might assume that these crows are yet another endangered species on its way to extinction. But that is not really the case. As a species, the house crow is doing very well in many parts of the world, including in its original "native" range in the Indian subcontinent and across into Myanmar and Yunnan Province, as well as in various other countries in which the species has subsequently arrived with a little, or a lot of, help from people (Nyari, Ryall, and Peterson 2006). In fact, it is precisely because this species is doing so well in these other places—perhaps too well—that in early 2014 Dutch authorities began to kill the small population of about thirty-five to forty house crows that had been living in Hoek van Holland for the past twenty years. Traps seem to have been ruled out early on—crows often don't fall for them—so shooting perched birds with an air rifle was adopted as the preferred approach. Crows quickly began falling to the ground.

I arrived in Hoek van Holland about four months into the killing, drawn by a desire to understand these birds and the government's decision to be rid of them. Despite their rarity—and the wariness that comes with persecution—I found some of the remaining crows quickly that morning. As I approached the train station where I had seen them the previous day—where I had, in fact, been shown them by a local resident and passionate crow observer named Sabine Rietkerk—a single crow swooped in low over my head and landed on the metal banister in front of me. I stood and watched the bird for a few moments until it flew a little farther on. I followed, climbing the concrete steps up a small embankment. There I sat on a bench for about fifteen minutes, watching this house crow and one other, along with about twenty jackdaws, as they collected food among the freshly cut grass.

When the house crows headed off, I followed, at a respectful distance. We headed in the direction of the water, a part of town I hadn't yet seen. One of the crows landed on a large sign by the edge of the waterway. As I turned to look at the bird I noticed my larger surroundings for the first time, my vision drawn across the water to the Port of Rotterdam. Before this moment I had only had a vague sense of this place. I knew this was a shipping area, that the Port of Rotterdam was Europe's largest port, and that all of the traffic to and from it passed along this stretch of water, but I had no idea of the extent of the port itself. Assuming it to be located miles away in the city of Rotterdam, I had not realized that it stretched for forty kilometers all the way to the coast—and, in fact, beyond, out into the North Sea.

Emerging out of this moment of encounter, this chapter is a reflection on these crows and this port. Working through the lens of "hospitality," I am interested in the ways in which other species are made welcome—or not—in the places that we call our own. The Port of Rotterdam situates my thinking. There is an obvious connection between these crows and this port: it was likely as stowaways on one of the many large ships coming in and out of the port that the crows made their way to the Netherlands in the first place. But my interest is primarily in a more diffuse set of connections between them. As one of the largest ports in the world, the Port of Rotterdam is an engine for the global patterns of production, trade, and consumption that are today ushering in what many are calling "the Anthropocene"—a proposed name for a new geological epoch, our current one, in which humanity is seen to be playing an increasingly significant role in the shaping of the planet, across interwoven bio-geo-chemical terrains.

With these larger planetary processes firmly in mind, the first two sections of this chapter explore the way in which the deadly treatment of crows and the massive expansion of this port might both be grounded in the same broad dynamic, specifically, in an assumed entitlement to the world that is today increasingly "authorized" by the ongoing transformation and destruction of environments. In particular, it is the circular, self-reinforcing (il)logic that binds together the destruction or "marking" and the appropriation of the world that interests me. The final section of this chapter works to imagine an alternative to this situation. In explicit contrast to the increasingly common calls for new forms of global environmental management, this section asks what other kinds of responses—to the world, to this place, to a small group of crows—might be open to us, might in fact be *demanded* of us, in our Anthropocene era. How might this site help us imagine ways of being with others outside of the appropriative logic of the host? Ultimately, this chapter seeks a more situated way into the relatively abstract notion of the Anthropocene: in Franklin Ginn's (2017) terms, it works to "emplace the planetary."

The declaration of the Anthropocene is explicitly about the *global* and the *long term*: anthropogenic marks inscribed in Earth's strata for countless generations to come. But this chapter argues that paying attention to the Anthropocene is about more than "zooming out" to larger temporal and spatial scales. This is certainly part of what the Anthropocene requires but by no means all of it. The Anthropocene's placetimes are more complex than this. As Michel Serres notes, "every historical era is . . . multitemporal, simultaneously drawing from the obsolete, the contemporary, and the futuristic. An object, a circumstance, is thus polychronic, multitemporal, and reveals a time that is gathered together, with multiple pleats" (Serres, Latour, and Lapidus 1995, 60; Bennett and Connolly 2012). As such, beyond simply zooming out, the Anthropocene, like any other era, demands that we pay attention to the way in which specific moments, specific situated circumstances—like these crows in this place at this time—are haunted, structured, and shaped by legacies of the past and the anticipation of what is to come. Imagined futures "are not merely geometrical extensions of time. They haunt our presents, obeying architectural laws that look more like Gaudí than Euclid" (Rosenberg and Harding 2005, 4). Drawing these threads together, this chapter asks about contemporary modes of gathering up times and places in Hoek van Holland. Which futures are being imagined or denied? How are connections being made between some times and places and not others? The Anthropocene is a particular mode of imagining, of summoning up, placetimes. What does it do, what *might* it do, to our hospitality?

UNWELCOME

The origins of the small population of house crows in Hoek van Holland are somewhat unclear but can probably be traced to the arrival of two birds around 1994, stowaways on board a cargo ship. No one knows where they came from. Some have speculated that it was probably from the Suez Canal in Egypt (Slaterus, Aarts, and Bremer 2009, 7) or perhaps from the large population in Aden, originally established by the British in the hopes that the birds might provide an effective cleaning service for the town (they didn't).[1] There are many possible origin sites to choose from. Since at least the 1800s, house crows have been traveling around the world by ship. In some cases they have been deliberately moved by people, but in many other instances they have practiced what biologists call "self-introduction" (Ryall 2003, 167). As a result, breeding populations of the species can now be found in roughly twenty countries outside of what is considered their native range (Ottens and Ryall 2003)

Wherever they live, house crows have a close association with people and our dwelling places, thriving on our waste and taking advantage of diverse urban and rural resources. This is so much the case that the species is now considered to be an "obligate human commensal" (Nyari et al. 2006) with no known populations living independently of us. Of course, house crows do better and worse, and live in different ways, within the diverse forms that human communities take around the world. But the fact remains that in each place that our two species cross paths, humans have (usually unintentionally) taken up the role of providing the things that house crows need, our waste becoming their primary food resource (Ryall 2003, 168). You might say that, in so far as these birds have a "natural environment," we're it. Although I have found no published explanation of their common name, it is presumably this close association that is captured in the English "house crow," the Dutch "*huiskraai*," and the many similar names in other languages.

With the exception of the small Dutch population, however, almost all other known house crows live in tropical and subtropical regions. It was initially assumed that the pair of birds in Hoek van Holland wouldn't survive the cold winters and certainly wouldn't be able to breed successfully. But they were able both to survive and to breed, albeit modestly. Over the intervening twenty years, two crows became about forty—perhaps slightly more, perhaps slightly fewer—no one knows because no one was really counting. The vast majority of these birds were likely descendants of the original pair, with the odd new arrival by ship helping to prevent a genetic bottleneck. These house crows settled in with the local jackdaws, sharing the same communal roosts at night. Quietly, they got on with their lives, finding new ways to live in this environment and new resources to exploit. Until, of course, all that changed in 2014.

The commencement of killing was not without some warning. After decades of peaceful coexistence, the government of the province of South Holland ordered a risk assessment of the population, published in 2009. The risk assessment didn't make any specific recommendations, but shortly after its release the government gave the order that the population be eradicated. This process was initially halted by a legal challenge from the Dutch animal-welfare group Faunabescherming, on the grounds that the species was officially protected as an "unassisted" arrival. In 2013, after the government removed the species from the protected list, the court threw out the legal challenge, ruling that the government had established that eradicating the population was necessary to prevent possible future damage to crops and wildlife.[2]

The term "future" is key here. As Sharon Dijksma, the minister responsible for the decision, is reported to have told the parliament: "taking action now will stop a lot of future problems" (*DutchNews* 2013). While not a single complaint had been made about these birds, some biologists and managers argued that if left unchecked the population might grow considerably larger and become a "pest." According to Colin Ryall, a British biologist who has studied house crows all over the world and provided input into the Hoek van Holland management effort, it is likely that the slow growth of the population up to this point had simply been a "lag phase" during which the birds were adapting to their new environment. In an interview he told me that in the years to come the population would likely have grown exponentially. The government's risk assessment combined this possibility of future growth with the impacts that large populations of house crows have had in some other parts of the world: "In many countries where the species has reached high densities it is seen as a pest that should be controlled or eradicated. Damage to crops, negative impact on native species and the spread of diseases have all been reported" (Slaterus, Aarts, and Bremer 2009, 7). It was these visions of the future, coupled with a general concern about nuisance—noise and shit—that were publicly used to explain the necessity of this killing.

But some activists and locals saw things differently. In an interview, Harm and Norman, the founders of Faunabescherming, told me that these potential impacts are a convenient justification for a politically motivated killing. In their view, the new Dutch agency responsible for controlling introduced species, the Food and Consumer Product Safety Authority (Voedsel- en Warenautoriteit), targeted this population and exaggerated the scale and likelihood of its potential impacts in an effort to demonstrate its own importance and efficacy. From this perspective, house crows were selected as the target—rather than one of the thousands of other introduced species in the Netherlands—for two reasons. First, pressure from the United Kingdom, worried that the crows might use the Netherlands as a base to spread out around Europe, and second, because there weren't too many house crows, the department had a chance for a quick, albeit largely symbolic, victory. Ironically, from this perspective it was the lack of growth of this population—its *lack* of invasive spread—that singled it out as a suitable target for eradication.

This line of argument explicitly works against dark visions of a Hitchcockian future overrun by crows. Alongside this alternative framing of motivations, critics from Faunabescherming and elsewhere have pointed out that no detailed study of these birds was carried out: for example, as the government's own risk assessment notes, their "nesting and roosting behavior remains remarkably little studied" (Slaterus, Aarts, and Bremer 2009). Nor was any monitoring conducted for potential effects on the surrounding biodiversity or agriculture. While these studies couldn't have provided a crystal ball, they would have been useful inputs into this process. We are drawn here into an appreciation of what Emily O'Gorman (2017) has called "imagined ecologies," in which ways of understanding—rightly or wrongly—a species and the (potential) interrelationships that constitute places come to matter profoundly. In the absence of local data, all that the risk assessment was able to do was point to studies in vastly different parts of the world, in terms of climate, competitors, and the ready availability of human waste. On this basis it was determined that there was a *good enough chance* that these birds would become a pest in the future and that if this were to happen it would then be too late to act. In short, the killing was presented as a precautionary response to an uncertain but likely dark future.

What does it mean to be declared to be "unwelcome" in this way? Jacques Derrida provides important insight here. While his notion of an "unlimited hospitality" has received a great deal of attention in animal studies and related fields (Oliver 2009, Lawlor 2007), his more general caution about the territorial work of hospitality is less often discussed. Again and again, Derrida reminds us that all acts of welcoming are grounded in acts of appropriation: "To dare to say welcome is perhaps to insinuate that one is at home here, that one knows what it means to be at home, and that at home one receives, invites, or offers hospitality, thus appropriating for oneself a place to welcome [*accueillir*] the other, or, worse, welcoming the other in order to appropriate for oneself a place" (Derrida 1999, 15). In short, to welcome is itself sometimes an act of appropriation, while in other contexts one is simply *able* to welcome because of a prior, more or less violent, appropriation (although perhaps these two moments are never so neatly divided). Although Derrida does not emphasize this fact, the act of *un*welcoming, of course, follows this same logic: *whether we welcome or not, to assume the power to do so is to appropriate to oneself a place*, it is to claim the right to decide who comes and who goes.[3]

Viewed in this way, the unwelcoming, and so killing, of these house crows can be seen to be grounded in an act of appropriation. This kind of response to populations of species who, for one reason or another, are taken not to "fit" within the confines of a particular place as it is imagined by those with the power to shape it is a common occurrence today. Of course, a place is always a placetime. As numerous theorists have noted, so-called introduced or newly arrived species are frequently understood to be out of place in this way, usually with reference to a specific, imagined, and idealized past ecosystem. But these house crows remind us that imagined *futures* are also important, arguably more important, in determining which of the literally thousands of "out-of-place" species will

FIGURE 3.2. Three house crows perched on the welcome sign in Hoek van Holland, the Netherlands.
Source: Photo courtesy of Garry Bakker.

actually be *targeted* for eradication, that is, which will be rendered lethally unwelcome.[4]

The arrival of a stranger is always haunted in this way, grounded in specific pasts and futures imagined and/or lived. As Heidrun Friese (2004, 70) notes, there is always an uncertainty "which penetrates and transcends [the potential guest's] arrival, his descent, origins, his 'domestic situation' is unknown and his intentions are opaque . . . it is precisely this association with uncanny danger which allows the rendering of 'the' stranger into a scapegoat (Rene Girard) or the public enemy." Clearly, which histories and futures are imagined and invoked matters profoundly for the stranger's reception.[5] In the particular case of these crows—strangers from unknown lands, arriving out of the blue on a foreign ship—dark visions of what might be gave rise to fears and preemptive actions. As Melinda Cooper (2006, 125) notes, preemption "transforms our generalized alertness into a real mobilizing force, compelling us to become the uncertain future we're most in thrall to." Dutch authorities imagined a future overrun by crows, and they said "you are not welcome." In doing so, however, they appropriated to themselves not only a place but the future itself. In the act of killing these crows, the Dutch authorities shut down other kinds of possibilities: that these birds might not have been pests, that we might have found other ways to live together, that the presence of crows might have become an invitation to some other, some more flourishing, future for everyone.[6]

ANTHROPOCENE ENGINE

It was on the morning of my second day in Hoek van Holland that I "discovered" the Port of Rotterdam. Immediately, my mind was occupied with thoughts of the entanglements of crows and ships. Of course, this port was their arrival point, but the more I reflected on it, the more I realized that taking this port seriously would be a vital condition for coming to terms with this little group of birds, with how it is that we ought to understand and respond to their presence in this Anthropocene era.

With this thought in mind, on the afternoon of that same day I boarded a ferry headed to a visitors' center in the Port of Rotterdam with the unlikely name of FutureLand. It was a windy and overcast day. The ferry traveled out into the thick of the port, where we were surrounded by factories and massive cargo ships. I got off at the first stop. I had been warned by the ferry operator that it was a long walk to FutureLand, but as he knew of no better public-transportation option, I went with it. As I disembarked I was immediately struck by the strangeness of this industrial landscape, in particular, its complete disconnection from anything approximating a human scale. This is a place designed around the needs of huge ships, trucks, and factories. Distances are vast, not quickly crossed on foot. For about an hour I walked along the highway watching two ships being simultaneously loaded and unloaded. As I went I counted dead birds on the side of the road, mostly seagulls—eleven in all—presumably hit by large trucks like the ones that now whizzed past me.

When I arrived at FutureLand it turned out not to be a visitors' center for the whole Port of Rotterdam but rather just for Maasvlakte 2, a massive extension of the port that has taken over six years to complete and will triple its capacity. In the absence of available land, the Port Authority

FIGURE 3.3. A factory in the Port of Rotterdam, the Netherlands. *Source*: Photo by author.

opted to build out into the North Sea, employing the same approach that has been put to use in other parts of the world, for example in the construction of Dubai's (in)famous Palm Islands.

I was expecting that there would be a lot of visitors—big ships have a strange effect on some people—but I was still surprised by the size of the crowd. There were probably 150 people in the relatively small space, with visitors coming and going the whole time, some taking the extended bus tour of the port. Within FutureLand there was an exhibition on containers through the ages, walls of brightly colored pictures and information, and several films and interactive exhibits, all explaining the importance of the port and documenting the ingenious, high-tech construction of the expansion.

I watched a film that followed a crane operator as he learned to place the massive stones that would form the foundation of the new land. These stones had to be precisely placed under the water, using sophisticated imaging equipment. Tons of sand were then dredged from the seabed off the coast, shipped back to the port, and sprayed in a slurry out into the water, eventually to build dry land reinforced by massive concrete structures. Another interactive exhibit allowed visitors to lie down and stare up into the guts of a model of the Earth—a large, hollowed-out globe suspended from the ceiling—to watch an animation about the importance of shipping to the Dutch and European economies and the vital role that the port plays (see figure 3.4). The audio announced: "From apples to cars, and from computers to raw materials for the chemical industry, virtually anything that can fit in a container is shipped by container. . . . And all those goods are just passing through, through Rotterdam, on their way to 500 million consumers in Europe. Plus one. And that's you!"[7] In short, FutureLand offered the visitor a celebration of trade and a testimony to the triumph of human ingenuity, in the form of Dutch engineering, in modifying this place to enable the future expansion of that trade.

There is, of course, another side to this place. In contrast to this celebratory presentation, we might understand the Port of Rotterdam as a key driver, an engine, of the Anthropocene. From climate change and the capture of the nitrogen cycle to large-scale landscape modification, mass extinction, and lingering toxic legacies, all of these proposed indicators of our new geological epoch are significantly tangled up with this place. All day and all night, seven days a week, ships from all over the world arrive and depart: this is how the old-growth forests of the world become floorboards, it is how we are able to keep burning fossil fuels at low prices. Global shipping networks are, of course, also how much of the world has been able to outsource its dirtiest industries, alongside its carbon-emissions responsibilities, to a growing constellation of "shadow places" out of sight and mind (Plumwood 2008). In addition to all this *movement* of goods, the port itself—occupying an incredible 105 square kilometers—is one of the world's largest oil and chemical centers: home to a broad range of refineries, factories, and other facilities taking advantage of the easy access to global markets.

Alongside these broad environmental impacts, of particular significance for the current discussion is the fact that international trade, and in particular shipping, is today also playing a central role in the unintentional global mass movement of biodiversity. In this context, house crows

FIGURE 3.4. An audio/visual presentation on global shipping at FutureLand, Port of Rotterdam, the Netherlands. *Source*: Photo by author.

are far from being alone. All over the world species are using shipping networks to travel. In fact, the degree of international trade in a country is now understood to be the most reliable predictor of the number of introduced species found there (Westphal et al. 2008). While many of these species probably cause no significant problems at all in their new homes, globally newly arrived species are one of the key causes of our current mass-extinction event (Davis et al. 2011). But, in a host of other ways too, international trade is today understood to be "driving" biodiversity loss (Lenzen et al. 2012).

To some extent, of course, my framing of the Port of Rotterdam as an "engine of the Anthropocene" is premised on a simplification. Rotterdam is just one port, albeit the largest one in Europe, one node in the dispersed network that is contemporary global trade and consumption. The Anthropocene isn't located in any one place; it doesn't "take place" here, or there, any more than anywhere else (a situation that, as numerous theorists have pointed out, is one of the many problems that we have in visualizing, representing, and responding to it).[8] Nor can the Anthropocene simply be reduced to contemporary processes of manufacturing, trade, and consumption, however important they are in its production. But some places do play a particularly profound role in flattening out and marking the world in the specific ways that will produce a new geological epoch. The Port of Rotterdam is one such place: not just a symbol but a manifestation and a vital material-semiotic enabler of our Anthropocene era. Paying attention to such sites is an essential part of the work of "situating" the Anthropocene: of creating attachment sites to explore its specific cultural, economic, philosophical, political, and technical contours in their always unequal forms.

At its core, the Anthropocene might be understood as an effort to draw attention to the current period as one in which humanity has profoundly *marked* planet Earth. Of course, there are often significant problems associated with attributing actions or characteristics to an amorphous "humanity," a point I will return to below. For now, however, I want to dwell with this central notion of the Anthropocene, a term that, in Ben Dibley's (2012, 139) words, "vividly captures the folding of the human into the air, into the sea, the soil." In contrast to Foucault's classic image of the end of "man," "erased, like a face drawn in sand at the edge of the sea," Bronislaw Szerszynski (2012, 180) notes that in the Anthropocene "the fate of 'man' . . . is not that he will be erased, but that he will be made immortal, as a trace preserved forever in the rock." From the perspective of stratigraphers, debating the formal adoption of the term, the Anthropocene is entirely about these anthropogenic marks, signs laid down in rock that might be visible to future geologists: new minerals, new strata (including new disruptions to old strata), new chemicals, new abundances and absences in the fossil record (Zalasiewicz et al. 2011). There is an odd practice of "speculative stratigraphy" at work here: imagining a future geologist, imagining future Earth strata, and taking a best guess at what they might say to each other. Like the imagined futures that ground the reception of the (potential) guest, the official declaration of the Anthropocene, with all of its consequences for how we will read our present (and past), is grounded in future relationships that we can only speculate about.

But from another vantage point, the particular marks of interest to geologists and stratigraphers potentially miss a great deal of what matters most about the Anthropocene. Whether or not the term is formally adopted by the scientific community, it is undeniable that the planet is currently undergoing a period of massive transformation. Many current changes will leave lasting scars, but many others may not. Whether these incredible marks end up being "good enough" for the purposes of hypothetical future geologists, doesn't necessary say anything at all about what kinds of changes they are for living beings, either now or into the future.[9] As a result, for a growing number of people, the reality of the Anthropocene is not something that will be decided by stratigraphers. This term has given "lithic stability" to many of the broader anxieties of our times (Sandilands 2018) and in so doing has become shorthand for our current more-than-ecological (but distinctly not just geological) predicament.[10]

My point in drawing attention to these anthropogenic marks is not to revisit the tired environmental refrain about how humanity has touched, stained, everything with our presence and that consequently there is no nature left. I don't believe in this kind of dualized "nature" or in this singular "humanity," and I have no interest in arguing for their resurrection/fabrication. Instead, I am interested in what the emphasis on marking at the heart of the Anthropocene discourse might mean, specifically, in what it is that these marks might be taken to *authorize*.

Michel Serres (2011) is my guide here, specifically his suggestion that we might rethink the history of property as one in which marking and staining are perhaps *the* central processes through which one lays claim to something. The person who spits in the soup, the fence builder or clearer of the forested land, and even labor theories of property (often rendered as the mixing of sweat with the land) all seem to conform to the notion that a mark, a polluting element, might be utilized in order to make a claim of ownership. As Serres notes, this practice also extends beyond the human species. Understood in the broadest sense, many other animals, including crows, also engage in forms of marking to appropriate a territory, to lay claim to a place: this might involve all manner of advertisement and alteration of the acoustic, olfactory, and visual terrain. I cannot follow Serres to the outermost edges of this argument: not all human cultures—let alone all species—"do" territory and appropriation in this way (and we certainly don't all agree on what might constitute a legitimate form of marking, a point that the violent history of *terra nullius* reinforces). Yet Serres does capture something important here. For many of us at least, "marking"—in one way or another—is an important, perhaps even commonsensical, mode of appropriation.

From this perspective, the Anthropocene takes on a new significance. We, at least some of us, have long marked and felt like the masters of this or that square of the Earth, but in our current time these markings—from toxic waste and mass extinction to climate change—seem to be being extended to the whole of the globe and into the distant future. It is, in no small part, the incredible *reach* of the actions of current generations that the designation of the Anthropocene, as a *geological* epoch, attempts to grasp and name. In addition to these real physical alterations, through the deployment of concepts like "Anthropocene" many of us are increasingly narrating our world as one thoroughly marked and refashioned by "humanity." Serres helps us see how this profound marking—material and discursive—might itself be a territorial claim of sorts, a claim to the Earth and its future.

Tellingly, many of the disputes over what to call our new geological epoch reproduce precisely this logic. They ask: "who is primarily responsible for these destructive marks?" That entity or category—be it Anthropos, capital, the economy—should have its name affixed to this placetime: thus suggestions that we may now be living in the "Capitalocene" (Moore

2014). However apt these suggested names might be, however much they might be intended as acts of apportioning responsibility, are they not simultaneously a kind of branding, a kind of appropriation of the earth and its future *for* humanity, or *for* capital? Perhaps an entirely new way of naming is in order, one in which the plunderer does not inherit the earth? Sadly, however, the ship seems to have sailed on "the Anthropocene."

But thinking with Serres about the Anthropocene also opens up the question of *who* it is that is marking and appropriating the world at our present time. While much of the Anthropocene rhetoric names "humanity" as the culprit, it is clear that some forms of humanity, some modes of social, cultural, political, and economic life, are far more central than others, principally those that have dominated since the Industrial Revolution and especially since the "Great Acceleration" of the 1950s (McNeill and Engelke 2016; Steffen, Crutzen, and McNeill 2007). Equally, it is clear that some humans will suffer—and many are already suffering—a great deal more than others as a result of the entangled processes of change that characterize the Anthropocenic Earth: while the Port of Rotterdam brings wealth to many, it is now also, in the words of a recent *New York Times* article, "Europe's main external garbage chute," a central node in the (often illegal) exportation of a range of toxic materials to poorer parts of the world (Rosenthal 2009).

As a result, in so far as it populates our discussion, the figure of the human needs to be read as a question, asking, again and again: *Which* human activities, *which* modes of organization, are implicated in contemporary processes of escalating environmental change? In the absence of such a questioning, the invocation of the Anthropos becomes an *appropriation* of the human as a figure and a possibility: the one who marks and claims the world *is* the properly human, rendered doubly so by the power to commit these acts and to define itself in the position of unmarked invisibility. In contrast, holding the human open as a question reminds us that an "alter humanity" is possible; indeed, it *already exists* in many placetimes.[11]

When the connection between marking and appropriation that Serres identifies is read through the figure of the "uncertain human," the Anthropocene can be understood as a period in which the dominance and authority of particular forms of human life is being *inscribed* in the landscape with an ever increasing intensity. This inscription *authorizes*, rendering invisible or unproblematic, an increasingly prominent role for (some of) humanity in the shaping of earthly futures. There is a deeply destructive and self-reinforcing dynamic at work here: the more Earth is transformed and marked, the greater the sense of entitlement to order and control it becomes.[12] Rhetoric is made strangely material: the Port of Rotterdam is rhetoric materialized, a site in which some human possibilities are inscribing their story into the Earth with a persuasive, self-authorizing force. It is in this expanded sense that I understand the Port of Rotterdam to be an "engine of the Anthropocene": a site of profound material consequence for contemporary processes of earthly change that simultaneously and insidiously authorizes its own existence and the escalation of its impacts.[13]

In this way, more and more of the world is today being appropriated by us—whoever "we" might be—seen to be, first and foremost, "our places": our cities, our farms, perhaps even our countries. We decide who belongs and who doesn't. This entitlement, the obviousness of this claim, makes us more and more unwilling to make room for others, to be inconvenienced by them, while simultaneously making us more and more oblivious to the effects of many of our own actions, as the real—or at least more significant—cause of our current ecological predicament. As Claire Colebrook (2012, 189) has succinctly put it: "Is not the notion that the earth is *our place* precisely what has blinded us to the ravages of our mode[s] of life?" Val Plumwood (1993, 193) called this appropriative attitude one of "mastery," in which all others are taken to exist to satisfy one's own needs and whims and are given purpose only in relation to the master's projects. Taken to its extreme, this attitude produces a "slave-world" in which "what can be incorporated into the empire of self is permitted to exist in assimilated form and what is not of use is eliminated."

Returning to the language of hospitality, it is this approach to the world that makes "us" the hosts and others, permanently, guests in our space, by our grace. From this perspective the phrase "Welcome to the Anthropocene"—the headline of the *Economist*'s issue on the topic—might be read as both deeply troubling and thoroughly appropriate. Just who is doing the welcoming here—humanity, the magazine, the particular economic community that it represents—and who is being welcomed remain characteristically unclear. What is clear, however, is that the act of welcoming is here coupled with the appropriation of the world. In the very first line of the leading article in the issue we are told: "The Earth is a big thing; if you divided it up evenly among its 7 billion [human] inhabitants, they would get almost 1 trillion tonnes each" (*Economist* 2011).

It was these appropriative dynamics that struck me so palpably at FutureLand. In contrast to visions of crows taking over Hoek van Holland (and perhaps Europe) and my own dark musings on the challenges of the Anthropocene, at FutureLand I encountered a celebration of what is to come, a celebration of the expansion of global trade. There is a profound disconnect at work here, between this place, its surroundings, and its consequences. While across the water, back in town, wildlife officials were busily killing crows, in the Port of Rotterdam the very processes of shipping that brought these birds and countless other new species to the area were being drastically scaled up, with all of their attendant local and global impacts. Here, at the epicenter of "our" remaking of the world to suit "our" designs and whims, the lives of forty crows could not be tolerated. In this massively transformed and transforming landscape we worry that *they* may one day harm local biodiversity, that *they* may one day become an inconvenience to *us*. This is the (il)logic of mastery at work.[14]

OUTSIDE HOSPITALITY

On my last evening in Hoek van Holland I set off in search of the house crows' roosting trees. Following the lines of flight of more conspicuous jackdaws, with whom I had been told the crows sometimes roosted, I arrived at a little stand of trees on the edge of town. The area was a hive of noise and activity as I approached. It was now mostly dark, but I could still make out the silhouetted movements of hundreds of corvids flittering between branches, squabbling and calling out as they went. All of the sounds were distinctly jackdaw, but a couple of times I thought I heard the harsher calls of a house crow. At one point a large group of birds took to the air together and left the trees. I wondered if my presence was disturbing them. Alongside the many house crows shot in the past months, some jackdaws had accidentally been shot too. Perhaps all of these corvids had become wary of humans hanging around under their roosts.[15]

I left the trees and headed back into the center of town, wondering as I walked about this space in which house crows and jackdaws are learning to make room for each other. While all corvids are territorial to some extent when nesting, their practices differ markedly. Breeding pairs of some species set up large territories and vigorously defend them from potential competitors or predators, but both jackdaws and house crows tend to nest communally, with many pairs in the same tree, each maintaining a relatively tiny "territory" around their nest. And, of course, when not breeding, both of these species of birds tend to roost communally at night (Goodwin 1986). This is precisely what they have done in Hoek van Holland, crafting a new space of interspecies roosting. In this zone of cohabitation—which is far from idyllic, far from risk free—perhaps we can find inspiration, even if not lessons, for alternative avenues into the future?

In large part, this chapter is an effort to retell the story of the house crows of Hoek van Holland in a way that attends to, and so might escape, the broader dynamics of the appropriation of the world and its future that lies at the heart of this Anthropocene era. To this end, the final section explores what it might have meant to respond to these crows differently, outside of the logic of the host. This approach is grounded in the

FIGURE 3.5. Five house crows in Hoek van Holland, with the Port of Rotterdam visible in the background. *Source*: Photo courtesy of Luuk Punt.

understanding that what the Anthropocene requires is not more or less hospitality or even a revisiting of whom we are hospitable to. Instead, what is needed is an entirely new frame of orientation. In contrast to the position of host—which no matter how hospitable is always still premised on an appropriative claim—this position would acknowledge that the world and its future were never ours (whoever we might be) to give in the first place, to welcome or not.[16]

The approach proposed here is in explicit opposition to the increasingly common calls for extended forms of "human" management of Earth. Grounded in a very similar logic to the one identified by Serres, these calls frequently move from the claim that "humanity" has marked and damaged the world, ushering in the Anthropocene, to an assumed right, or perhaps an obligation, to take up new and expanded forms of control (Crist 2013b). This easy slippage was already clear at the founding moment of the Anthropocene discourse, in the article by Crutzen and Stoermer (2000) that launched the term onto the global stage. Having diagnosed this new geological epoch, the authors conclude with the following words: "An exciting but also difficult, and daunting task lies ahead of the global research and engineering community to guide [hu]mankind toward global, sustainable, environmental management" (18).

In the intervening years, this same logic has been deployed in a range of different ways. For some proponents of global management, the profound contemporary transformation of the world is read as an indication of human ingenuity and skill (even if not always perfectly realized). In this context global management is a case of humanity assuming its rightful place and taking up the Baconian mandate—or is it an Edenic decree?—to produce a "garden planet" in which nature is put to work to sate "our" ever growing resource needs (Kareiva, Lalasz, and Marvier 2011).[17] For others, the marks made by humanity on the Earth are read as indicators of *desecration*, yet these accounts often end up in much the same place: demanding that humanity take responsibility for the damage that we have caused through the careful and close management of the Earth; this is control and appropriation after another fashion. For most commentators, the situation seems to lie somewhere between these two poles. We are told by Erle Ellis (2011, 44), for example, that: "Our powers may yet exceed our ability to manage them, but there is no alternative except to shoulder the mantle of planetary stewardship. A good, or at least a better, Anthropocene is within our grasp." Or, in Stewart Brand's terms: "we are as gods and have to get good at it . . . [this] involves what ecologists call ecosystem engineering. Beavers do it, earthworms do it. They don't usually do it at a planetary scale. We have to do it at a planetary scale" (Brockman 2009).

In this way, the Anthropocene discourse is being enrolled into projects of ever-increasing control and management: the scale of the impacts, the marks, demands an equivalent grandiosity in terms of mastery, and so we get projects to geoengineer the climate and remake Earth's prehistoric ecosystems. These are projects grounded in what Gerda Roelvink and Magdalena Zolkos (2015, 47) have called a "hyper-humanism which seeks to manage and ultimately master the ecological crisis." While these projects might, in principle, take a wide range of forms, in practice they are increasingly being imagined in alignment with dominant neoliberal agendas, in the work of "ecomodernists" and others. Far from challenging attitudes of mastery, these approaches double down on them, reinforcing the foundational appropriative assumption that the world is, first and foremost, for "us." And so, the massive reordering of the planet envisioned by people like Michael Schellenberger and Ted Nordhaus from the Breakthrough Institute is one that is, as Clive Hamilton (2014) has succinctly put it, "system-compatible." While humanity might need to change its ways a little, the key focus is on reengineering the world to make it work for us. Hamilton again: "The answer, they say, is not to change course but to more tightly 'embrace human power, technology and the larger process of modernization.' " From this perspective there are no real biophysical limits to earth systems: by embracing new technologies and the market, we will find ways to overcome problems and continue to grow economies. Most of these solutions, it seems, will lie in increased intensification and, to borrow a term from Dr Seuss's *The Lorax* (1999), the "biggering" of urban environments, agriculture, aquaculture, energy generation, and fresh-water production (through nuclear power and desalination). In this way, more and more of the Earth can be put to work to satisfy the needs of a growing human population.[18]

Importantly, these calls for greater management of the globe also tend to invoke "the human" in a troublingly amorphous way: it is humanity that is to take up this mandate, burden, or responsibility of management, not the United States or some corporate entity, not climatologists or ecomodernists. This is the preferred framing precisely because it taps into a deep history in the West and elsewhere in which humanity has been seen to legitimately occupy a position of superiority in relation to the rest of the world: from the Great Chain of Being through Renaissance notions of the "dignity of man" to contemporary accounts of our unique genetic and creative capacities. In this way, referencing the human draws on a sense of "natural entitlement," falling easily into a storyline in which humanity assumes its proper place in the world. At the same time, however, referencing the human—perhaps strategically—obscures other pressing questions about precisely *who* it is that will take up these management roles, and to whose benefit. Humanity is complex and multifaceted; inevitably, some will benefit and others will not. Equally, however, in a time of mass extinction and the proliferation of factory farms, to name just two key issues, human lives are very definitely not the only ones at stake. In short, the human is at once too specific and too vague as a ground for imagined futures: who gets to take on the role of landscape designer and who is deemed to be a weed will matter profoundly on a garden planet.

Advocating for an understanding of the world outside of this logic of the host does not mean ignoring or doing nothing about the pressing environmental and social challenges of the Anthropocene. Nor does it require us simply to accept all changes to ecosystems, to refuse to take up any deliberate role whatsoever in their ordering. We are all inescapably entangled, *involved*, in the shaping of worlds. As such, not only are there not just two choices here—complete management or withdrawal—but neither of these extreme positions is really open to any fleshly, mortal being in the first place. Here we are reminded that what is at stake in efforts to manage the world is not an actual mastery but its pretense: "we" are not now, nor will we ever be, in control of the Earth (Clark 2011). In this light, the abdication of the position of the host is about humility; it is about accepting that the world is not ours to "sort out," to order unilaterally to a particular vision of how it "should" be. But as earthly inhabitants like any others, we must still alter places and have impacts and ideas about how things might be. What is required is that we take these up outside of a logic of *entitlement*, of assumed authority and legitimacy, in which others—human and not—must make all of the sacrifices while their own visions and needs remain permanently secondary. What is demanded, in short, is the cultivation of new approaches and sensibilities for crafting multispecies communities on an Earth that is *for* all of its living—and other-than-living—forms. This requires learning to ask a broader set of questions of ourselves and others, to hold open a broader set of possibilities and responsibilities.

From this perspective, hospitality reveals itself to be particularly unsuitable as a basis for responding to others in multispecies worlds (at least in the particular cultural form that this term takes in the work of Derrida and others). These are worlds in which lives are lived in zones of inescapable overlap. My house, my body, are always already others' territories too, often without us even really knowing about the others' existence. Not only are these overlaps not always detrimental, but many of them are necessary conditions for earthly survival (McFall-Ngai et al. 2013, Lorimer 2016). The notion that any place is exclusively or even first and foremost "ours" is not a helpful foundation for the crafting of multispecies community. The good of any one cannot be secured *against* all others (van Dooren and Despret 2018). The radical and fundamental limitations of hospitality—which are always constitutively entangled with complex temporalities—push us toward alter-territorialities. There are countless ways of doing and sharing space with others: alter-territorialities are possible—again, they already exist, both around the "weedy" (Tsing 2005) edges of empire and tucked away within its inner folds. These alternative approaches are not about welcoming others into one's own sovereign placetimes but about imagining and crafting less appropriative possibilities *with* others.

In Hoek van Holland this might have meant getting better at compromise: learning to live with crow shit and noise, even tolerating reduced harvests. Or, if crows did become too numerous in farmers' fields, it might have meant adopting the kinds of nonlethal methods of deterrence employed to deal with these birds in parts of India. Reduced harvests are more difficult to cope with in some parts of the world than others. In a place like the Netherlands such a response is well and truly possible.[19] Perhaps these impacts are some of the "prices" of a port—especially in an area and a country where much of the wealth (for many) has been built on hundreds of years of transnational openness in the form of shipping, almost all of which has flowed through Rotterdam. We don't get to have the world precisely as we want it, satisfying all our needs and requiring others to bear all the costs. What is needed here is a willingness to support or at least tolerate other species' own experiments in emergent forms of life for difficult times, experiments that will sometimes make us uncomfortable.[20] Of course, some people will always bear the brunt of this discomfort, and our modes of response and accountability must take these inequities seriously.

But alongside these concerns about agricultural impacts and nuisance, some conservationists worried about the potential future impact of house crows on local biodiversity. Might a substantially larger population of these crows—who, unlike many other corvids, work en masse to predate and raid nests—have wiped out much of the local avifauna? As Colin Ryall put it to me, the impact of house crows on native avifauna "has been devastating wherever they have proliferated." Might the more aggressive house crow even have eventually displaced or greatly reduced populations of European corvids like jackdaws and carrion crows (*Corvus corone*)? Ultimately, I am not convinced that enough time and resources were invested in asking these questions seriously. Killing emerged as a management option and was taken up too easily. But, even if such impacts were likely, much remains to be done to determine whether and in what form action might be warranted. Although it is frequently simply assumed that the changes brought about by newly arrived species are "harms," outside of an outdated vision of a static, stable, natural environment, this is something that needs actually to be shown. In a period increasingly characterized by strange new "recombinant ecologies" (Soulé 1990), produced in large part by the twinned and twined processes of globalization and climate change, such an understanding becomes increasingly unrealistic and problematic (Hinchliffe and Whatmore 2006).

At their heart, these are fundamentally ethicopolitical questions: questions about how we constitute multispecies communities, on what terms and through what procedures. These are topics that I explored in more detail in chapter 1, advocating for ongoing multiplicitous experiments in shared life. In Hoek van Holland it is certainly possible that such a process might ultimately have been unable to produce livable outcomes for people, house crows, and the many other species that together call this place home and that the best available course of action might have been to kill these house crows. I want to leave this question open; I do not have the information necessary to answer it well. Instead, my primary aim here is to dwell for a moment with the way in which, outside the logic of the host, killing crows—even if perhaps necessary—becomes an insufficient, an *inadequate*, mode of responsibility.

If the world is not first and foremost "for us," then we are required to think more expansively about the nature of "problems" like these house crows. For example, the huge toll that intensified agriculture, habitat loss, and ongoing hunting have taken on avian diversity in the Netherlands and Europe more generally (EEA 2015) needs to be part of this conversation. If house crows are a potential threat to avian diversity, then what about these other factors, which have consistently been found to be the key problems? In fact, according to BirdLife International (2017), poor agricultural land-management practices alone have led to an incredible 50 percent reduction in European farmland bird populations since the 1980s. Rejecting the position of the host requires us to ask about these practices, to insist that patterns of landscape transformation, no matter how profitable for some, cannot be allowed to be taken for granted as the assumed and unchangeable background against which to manage the potential biodiversity "crisis" posed by a newly arrived species like the house crow.

This approach will surely be dismissed by many. In times like these, we need practical solutions to problems. But when we allow issues to be defined by what is "practical" (according to whom?) or by the remit of the government department that has been tasked with them (who deal with wildlife, not agricultural policy or international trade), we severely limit our understanding and our options. Plumwood (2009, 113) called this the "tyranny of narrow focus and minimum rethink" and explicitly linked it to a mentality of mastery. She pointed out that "rethink deficit strategies do not encourage us to question the big framework narratives that underpin our extravagant demands or the associated commodity cult of economic growth, or to question our right, as masters of the universe, to lay waste to the earth to maintain this cult's extreme lifestyle." Only ever looking for immediately practical and achievable solutions means never being required to interrogate the broader underlying dynamics that generate these crises in the first place. In short, these "realist" assumptions all too often ground and legitimate worldviews of mastery that authorize ongoing catastrophe for others.

The point of refusing to take up the management of the globe is not to abdicate any responsibility for its shaping but rather to learn to craft new, less lethal and appropriative *forms* of responsibility inside thick place-times. The disavowal of a long and ongoing history of human impacts in this place, the ready anticipation of future "crow damage" and acceptance of a lethal response to it, the celebration of global consumption and trade and the casual denial of their consequences: again and again we see an unwillingness to make the *uncomfortable* connections that would demand more of all of us, analytically and ethically. To reduce the question of responsibility for the presence of these crows to the act of killing them is quite simply irresponsible, an irresponsibility made palpable in the shadow of the massive, expanding presence of the Port of Rotterdam.[21]

Outside of a logic of hospitality, on a world that is for all of its diverse creatures, futures must be constructed by fragile collectives that cannot claim absolute knowledge, power, or authority. Accepting this understanding might create an opening into different kinds of approaches to effecting change for the better, approaches that hold open the question of who acts and to whose benefit; are self-reflexive and demanding, interrogating deeper causes and possible responses, rather than rushing to "fix" others; that are collaborative and respectful of others' projects, visions for the future, and efforts to get on in a challenging world; and that relentlessly connect up issues with broader frames, challenging assumed and often easy categories and tactics.

Ultimately, I can't be certain that such an approach would have changed the fate of this small population of crows. What I do know, however, is that the response that was taken both manifests and reinforces the dangerous illogic driving our current predicament. It was a response inadequate to the challenges of our time. Holding these "two" stories together—that of the house crows and the Port of Rotterdam—is important work for a different kind of human-wildlife interface, one that we can no longer call "wildlife management." If the sudden discovery that we have blundered into a new geological epoch has taught us anything, it is that we cannot decide how the world will be. We do not know. Owning up to this uncertainty, to this radical *inability* to shape the world to (any of) our wills and designs, is a key part of undoing the pretense of mastery. If we want to have any chance at all of adequately responding to others in this Anthropocene era, then the crows of Hoek van Holland, and countless other species like them, must call us into a deeper sense of humility and compromise, a more expansive questioning and set of responsibilities.

Over the years since my visit to Hoek van Holland in 2014, I have remained in contact with Sabine Rietkerk, the crow supporter who first introduced me to these beautiful birds by the local train station. In August 2015, she informed me that five house crows were still alive, undoing the government's hopes for a quick victory. But slowly and steadily they disappeared, nonetheless. The hired hunter was able to kill most of them before the permit expired in 2017. At least one bird hung on, though. Sabine told me that this remaining bird was a cautious one, always on the lookout and quick to sound the alarm. Despite this fact, he was also the tamest of the birds in his interactions with her, eating from her hand and sometimes sitting quietly beside her on a bench. "I do not think he was shot," she told me. He was still around after the end of the shooting. "I think and hope that he died a natural death."

FUMIGATING

Crows fumigate. This is an odd and admittedly speculative proposition. It is, however, one too enticing to leave unexplored. Over the last few years I have come across several scattered references to interactions between crows and smoke in which the former seem to be deliberately bathing in the latter. After observing this behavior in Ireland, one person wrote to the Royal Society for the Protection of Birds (RSPB) to ask for an explanation. It seems that crows had been seen standing right alongside a chimney, preening themselves in the smoke (Hayward 2006). Why would they do such a thing? We don't know for sure, but a likely explanation, and the one offered in this case, is that "the smoke from chimneys does a great job at clearing irritating parasites, such as mites and ticks, from the birds' skin and feathers."

Certainly, birds have developed a range of approaches to fumigation—which is nothing more or less than the work of rendering a place, even if that place is one's own body, *unlivable* for particular others. Many birds dust bathe, a pretty self-explanatory and widely observed practice that maintains feather quality and may also help remove parasites. Some birds engage in a fascinating behavior called "anting," in which they either lie down on an ant nest or pick up ants in their beaks and rub them on their feathers. Either way,

---

FIGURE V.4.1. Illustration by Kirsty Yeomans (www.crowartist.co.uk).

the result is the same: the ants secrete chemical compounds such as formic acid that function as insecticides and likely help control lice and other parasites. Meanwhile, in the nest, some birds deploy botanical strategies, bringing in aromatic vegetation that may have antiparasitic properties (Suárez-Rodríguez, López-Rull, and Garcia 2013). But at least some crows seem to have developed this other approach—a pyrotechnique—for fumigation.

In addition to chimneys, a couple of brief reports have surfaced of rooks (*Corvus frugilegus*), again in the United Kingdom, making use of a now even more ubiquitous source of smoke in our cities: cigarettes. In 2007, a commuter at a train station in Devon reported seeing a rook pick up a recently discarded cigarette butt, move it from the tracks onto the platform, and open its wings over the still-smoking butt. A few minutes later, the proceedings repeated themselves with another rook—or perhaps it was the same one?—collecting another discarded cigarette butt (*Telegraph* 2007). The same behavior has also been observed among rooks in Ireland (Viney 2005), and a myth (unsubstantiated it seems) has even developed that a cigarette butt collected by a rook was the cause of a fire that damaged the thatch roof of Anne Hathaway's historic cottage in Stratford-upon-Avon (Wood 2015). I like to think that similar fumigatory aspirations lay behind the widely reported theft and distribution of a packet of cigarettes by a group of crows in the Maldives in 2010 (*Telegraph* 2010).

As far as I know, there haven't been any scientific studies of this behavior. If crows are using cigarettes in this way, though, it would seem to be a pretty clever development. Nicotine has long been used by humans as an effective insecticide—so effective, in fact, that closely related neonicotinoid pesticides are now thought to be playing a key role in the global decline of honeybees (Fairbrother et al. 2014, Green and Ginn 2014). The efficacy of nicotine in controlling pests can also be seen in cases in which urban birds have taken to opening up cigarette butts and including their fibrous contents in the lining of their nests. Preliminary studies of these birds indicate that this practice is effective in reducing parasites on growing chicks (Suárez-Rodríguez et al. 2013). Read in this light, crows fumigating themselves with smoke from cigarettes is perhaps just one more chapter in the world-forming, multispecies history of tobacco.

At the same time, however, it is a practice that twists and challenges us in interesting ways. How can crows deploying fire and smoke for their own purposes *not* threaten that archetypal image of Man, set apart from Nature by his mastery of the environment, his deployment of tools, his taming of flame? This is a well-told story. As a recent *Smithsonian* article puts it: "wherever humans have gone in the world, they have carried with them two things, language and fire" (Adler 2013). Charles Darwin ([1871] 1981, 137) made a similar connection, positioning fire as "probably the greatest [discovery], excepting language, ever made by man." The significance of this connection between Man and Fire has been explained in many ways, with use of the latter often seen to play a key role in the emergence of the former. As Nigel Clark and Kathryn Yusoff (2014) note: "Some evolutionary anthropologists have suggested that—both culturally and biologically—learning to handle fire is the single most important moment in becoming human." Another recent manifestation of this line of argument can be found in the controversial "cooking hypothesis," which positions fire as pivotal in the evolution of *Homo sapiens* through the way in which cooking provided access to the additional caloric energy required to fuel growing hominid brains (Wrangham 2010).

Of course, these stories of *Homo ignis*—hominid becoming Human through Fire—are not the only ones that can be told here.[1] An anthro-crow-centric approach to combustion thinking might begin with the iconic image of a crow lighting Tippi Hedren's cigarette during the filming of *The Birds* (see figure V.4.2).[2] This film has undoubtedly been a public-relations nightmare for the avian world, perhaps especially for crows and ravens, who take up a particularly menacing role in it. Again and again in my casual conversations with people who claim to hate crows, this film comes up as some sort of an explanation for the reasonableness of such a view. To my reading, this film is particularly fascinating for the way in which it unsettles anthropocentric assumptions, undermining human security through the familiar and the everyday (or, at least, the very white, 1960s, small-town American version of the everyday depicted in the film). There are no aliens or monsters from another world, just birds, inexplicably(?) turned on people. I wonder if this cigarette-lighting image was intended to soften, or perhaps just to complicate, the threatening relationships crafted in the film. It is an oddly domestic, even intimate, encounter between woman and crow. But read from a slightly mutated perspective, this image might offer an even greater challenge to anthropocentric perspectives than the

FIGURE V.4.2. Tippi Hedren having her cigarette lit by a crow. A publicity shot for Alfred Hitchcock's *The Birds* (1963). *Source*: Photograph © Philippe Halsman / Magnum Photos.

film itself. Read against the grain of familiar stories of a self-making Man who tames Fire, read as a representation of an actual (even if in this case staged) relationship between corvid and fire rather than a comical anthropomorphism, this image is a reminder that crows, and perhaps other nonhumans, also "do fire," deploying it to their own ends.[3]

What is at stake in these corvid interactions with fire and its remnants? What do rooks holding their wings open over cigarettes or chimneys have to tell us about the world? Are we perhaps here witnessing the birth of *Corvus ignis*? It seems to me more likely that fumigating crows are yet one more example of the way in which the world's many wakeful beings exceed any simple, singular narratives about where anyone has come from or is heading.

FIGURE 4.1. Raven in Joshua Tree. *Source*: Photo by author.

Chapter Four

RECOGNIZING RAVENS

Becoming Subject to Each Other

MOJAVE DESERT, UNITED STATES

"I love that photo. I'm thinking if you had the shell triggered so that at that moment, when the raven's face is right up close, it just went *bam!*, that bird would never touch another tortoise" (see figure 4.2).

I was sitting at a kitchen table with two biologists in the small town of Joshua Tree, California, in the Mojave Desert. The biologists, William (Bill) Boarman and Tim Shields, were telling me about the many creative projects they currently had underway that aimed to reduce the impact that predation by common ravens (*Corvus corax*) was having on threatened Mojave Desert tortoises (*Gopherus agassizii*). At this particular point in our conversation, Tim was explaining to me their plans to "weaponize" 3D-printed tortoise shells in an effort to communicate clearly and forcefully to ravens that tortoises ought not to be eaten. Contrary to what one might imagine on first hearing about this plan—an imagination well and truly reinforced by long histories of wildlife-management practice all over the world—this was not to be a lethal encounter. The exploding tortoise was not designed to kill, or even harm, but rather to draw ravens into alternative modes of recognition: asking them to see and to be in new ways.

The plan was relatively simple. Through the generous collaboration and 3D-printing expertise of Autodesk and Think2Thing, Tim and Bill had

FIGURE 4.2. A common raven (*Corvus corax*) exploring a techno-tort in the Mojave Desert, California (taken by a camera trap). *Source*: Photo courtesy of Bill Boarman.

already been able to produce lifelike tortoise shells (see figure 4.3). While Bill and others had made Styrofoam shell replicas in the past for research purposes, this new breed of "techno-tort" was uncannily realistic—realistic enough to fool me. I was told it could also fool biologists. But the real test was whether it would fool the ravens. In this regard too, early field trials seemed to indicate that the techno-tort was a success. Tim and Bill had placed nineteen of the shells out in the desert with motion-sensor trail cameras to observe any interactions with them. As we sat and talked that day they showed me still images—like figure 4.2—and film captured by these cameras. Of course, it is very hard to know what a raven is thinking as she approaches one of these shells. Is she curious about a strange, unfamiliar object in the environment? Is she wondering where this bad replica of a tortoise shell came from? Or does she believe that she has come across the familiar form that we call "tortoise" (and that she presumably knows and names in some other way)?[1] In at least some of the footage, there was a strong indication of a genuine experience of tortoise (mis)recognition.

FIGURE 4.3. A 3D-printed tortoise shell, or techno-tort. *Source*: Photo by author.

In one memorable clip, an adult raven confidently approached the shell and, without hesitation, flipped it onto its back and pounded down on the plastron (underside) of the shell with its thick beak (a relatively common mode of "opening" the still-soft shells of juvenile tortoises among ravens in the area).

While more research is needed, this initial interest from the ravens lays the foundation for the next stage of Tim and Bill's plan: the "weaponizing." In this phase, shells will be fitted with accelerometers so that when a raven flips one over, or otherwise disturbs it, the shell will be triggered to let out a spray of methyl anthranilate. This substance is not toxic; in fact, it is the chemical name for a compound familiar to many of us as artificial grape flavoring. It is, however, a substance that ravens and many other birds hate (and one commonly recommended as a bird deterrent). Tim and Bill are also considering a more low-tech option for outfitting the shells: packing them full of meat treated with another nontoxic substance that will, however, make a raven nauseated for a period. The hope is that this experience, frightful or nauseating, will provide the basis for a new and lasting perception of tortoises as distinctly inedible, that is, that it will encourage ravens to recognize tortoises differently.[2]

This chapter is an exploration of this and other efforts to intervene in human-raven-tortoise relations in the Mojave Desert, centering on the work of Hardshell Labs, a small company started by Tim that is something of a cross between a conservation NGO and a Silicon Valley–esque startup. Alongside exploding tortoise shells, their plans to reduce raven predation of tortoises involve lasers, drones, and even video games. At the heart of my interest in these projects is the question of recognition. How do ravens make sense of devices like techno-torts? How do people make sense of raven sense-making, and how are our ways of doing so bound up with the ways in which we recognize and understand ravens and their activities? In short, how do we appear to each other, and with what consequences for whom?

Drawing on a range of biological and philosophical literatures, this chapter explores recognition as a mode of opening into a world inhabited by and with others and being remade in the process. As will become clear, I do not believe that recognition is irredeemably tied to the particular humanist legacy of Hegel and others; this concept can be taken up, it can be inherited, otherwise. Nor is the concept of recognition that I have in mind here the orchestrated, one-sided, and frequently tokenistic one that has been the subject of sustained critique by Indigenous and postcolonial scholars in recent decades (Coulthard 2014, Birch 2018). While they are often far from innocent and equitable, at least some of the possibilities of recognition explored in this chapter open into spaces of *mutual* transformation and obligation that might just offer some fruitful avenues for getting on well together.

At a fundamental level, I take recognition to be a question of what it means to live well with others *as other subjects*: to understand ourselves as inhabitants of a world composed of diverse wakeful beings who are agentive, capable, meaningful, mindful, and attentive in some ways and not others. While subject/object dualisms are today frequently being questioned and, with them, the meaningfulness of these categories, at the heart of this chapter is a commitment to and an effort to articulate the importance of paying attention to this thing called "subjectivity," not in the abstract but in the specific forms that it takes inside relationships—its becomings—which are always rich with consequences, possibilities, and responsibilities.

This chapter begins with a brief background to the emergence of the "raven problem" in the Mojave before offering a fuller account of the current projects being developed by Hardshell Labs. The core of the chapter explores the way in which these projects function as "diplomatic" proposals, grounded in an understanding of ravens as wily, intelligent, and adaptive beings. Recognizing ravens as subjects in this way opens up a range of new possibilities for cohabitation, for emergent and imagined ecologies (Kirksey 2015, O'Gorman 2017). But at the same time, it also raises the specter of possible new forms of psychic and social trauma. The final section of this chapter focuses on the difficult, potentially traumatic possibilities that reside within these Hardshell Labs proposals: what new obligations might also emerge as we come to recognize others as subjects capable of—or susceptible to—distinctive forms of harm? What might we all become together, subject(ed) to these particular modes of and possibilities for recognition?

THE "RAVEN PROBLEM"

The decline of the desert tortoise is a fascinating and complex, not to mention tragic, story. While these incredible creatures are undoubtedly in need of more of their own storytellers, my focus in this chapter is a corvid one, or, rather, it is on those sticky sites of human-raven encounter that are shaped by a desire to conserve the tortoise. Tortoises have been known to be in decline in the Mojave Desert for several decades, being first listed under the Endangered Species Act in 1990. The causes of this decline are many. The Mojave is a vast desert that covers large areas of southeastern California and southern Nevada, as well as small sections of Arizona and Utah. Large areas of the desert are today protected, including Joshua Tree National Park, Death Valley National Park, and the Mojave National Preserve. But other significant areas have become increasingly densely occupied and crisscrossed by highways, from Palmdale in the west to the sprawling city of Las Vegas in the east. Long a home to Native American peoples like the Chemehuevi, Cahuilla, Serrano, Timbisha Shoshone, and many other tribes including, of course, the Mohave themselves, the area is today being transformed by human activities on an immense scale.

The causes of tortoise decline in the desert are multiple and complex. They are losing habitat to roads, to urban, military, and agricultural development, and to a range of industries, including the massive solar- and wind-power facilities that now dot the desert. One such facility, a solar thermal power plant in the Ivanpah Valley on the California/Nevada border, is thought to be the largest of its kind in the world, composed of over 170,000 mirrors and covering an incredible 3,500 acres of former tortoise habitat, from which tortoises were relocated during the construction (Garthwaite 2013). Meanwhile, introduced plants that have spread across the desert with the assistance of off-road vehicle recreation and cattle grazing have diminished the quality of the food available to tortoises while also making them vulnerable to increased and more intense wildfires. The Mojave's tortoises are also subject to accidental violence (for example, in the form of encounters with off-road vehicles) and occasional deliberate killing, and today they are thought to be being weakened by a respiratory disease that has perhaps been exacerbated by growing numbers of pet tortoises being released into the desert.[3] For a species that takes a long time to reach sexual maturity, these changes have been too much to bear.

But, of course, ravens are also an important part of this story of tortoise decline. Raven predation was first recognized as a significant problem in the mid-1980s, when scientists started finding empty, often punctured, tortoise shells under raven nests. In one frequently referenced example, 250 tortoise shells were found over a four-year period, scattered around the base of a single Joshua tree just outside one of the study plots where tortoise numbers were being monitored. Here and elsewhere in the desert populations of older, larger tortoises are in decline, but they are not the key cause for concern. Many of these tortoises are reproducing successfully, but while their offspring hatch, they never seem to become adults. They don't make it through the juvenile stage in their development—the first five to seven years of life, during which their shells are still soft and vulnerable to raven beaks—to take the place of older, reproductively active members of the population. While the wide array of threats to tortoises is putting pressure on animals of all ages, the raven threat is a relatively acute but particularly worrying one, focused as it is on this specific life stage.

Ravens and tortoises have both long inhabited the Mojave. In recent decades, however, things have changed. Many of the landscape transformations that have been so negative for tortoises have instead presented themselves as wonderful opportunities for these highly adaptive, generalist, and intelligent corvids. Growing human settlements have provided

FIGURE 4.4. Tortoise sign. *Source*: Photo by author.

year-round access to water and food in towns but also at waste facilities, farms, and along roads, in the form of a steady supply of roadkill. Where once raven numbers were restricted by harsh summers and winters, these limitations have been effectively lifted. Equally as importantly, urban development has provided perching and nesting sites where once the desert afforded none. While in some places Joshua trees offered ravens nesting sites, across other vast areas of desert the absence of trees meant that these birds were at best infrequent visitors and certainly not permanent residents. Now, power poles spread out into the desert in seemingly endless networks providing convenient nesting spots and hunting perches every four hundred meters or so. With these changes the raven population has grown dramatically; birds are now both living at higher densities and spreading farther into the desert. In the western Mojave, for example, raven numbers are thought to have increased sevenfold in the period from 1969 to 2004 (USFWS 2008, 4). Tortoises might be an important source of food for individual ravens in a few remote areas, but for the most part their deaths are probably just "collateral damage" in the life of a growing corvid population sustained by huge quantities of anthropogenic resources.

Nobody thinks that ravens are the only problem for tortoise conservation. Most people don't even think they're the biggest problem. They are, however, *a* key problem. The first management efforts centered on killing, usually referred to as "lethal control" (a terminology that I have deliberately avoided).[4] In 1989 and 1990, managers shot and tried to poison ravens in the area until a restraining order was filed by the Humane Society of the United States.

The Humane Society's primary concerns were that birds not responsible for taking tortoises would be killed and that other species of animals may be harmed by ingesting the avicide. The lawsuit was subsequently settled out of court, and the pilot program was terminated (Boarman 1993, 4). Building on this experience, in the early 1990s a small project was established to shoot "known offenders," specifically targeting ravens who "were found near a nest under which a minimum of three shells were found that showed evidence of predation."[5] Only about fifty ravens were killed in this two-year project. Surveying to identify "guilty" ravens was time consuming and expensive; shooting them was no easy task either, even for the trained professionals, some of whom are former military marksmen, who undertake this work on behalf of Wildlife Services (colloquially referred to as "gopher chokers").[6] When one raven was shot, their partner would often disperse, frequently getting away. That bird might subsequently be much more wary of people, even specifically avoiding the shooter and their car.

Over the intervening years, the killing of ravens in the Californian Mojave has taken place on a small scale, focusing on guilty individuals in this way. This approach is more difficult and expensive, but it has reduced conflict with animal-welfare groups.[7] While there is a clear appetite for expanding these programs into large-scale killing among some members of the conservation and wildlife-management community, many others oppose such a move, often citing the general lack of efficacy of such an approach. As Roy Averill-Murray, the Desert Tortoise Recovery Coordinator for the U.S. Fish and Wildlife Service (USFWS), explained to me: "From the research that has been done on predator control in general, including with corvids, in situations where you have such a huge subsidized population, the degree of effort to knock down the raven population as a whole in the Southwestern U.S., or just in southern California . . . it would be a massive undertaking. Only to have the effect that that number of ravens move in the next year. So, I don't see how it accomplishes anything. It might provide a little breathing room for a cohort of tortoises to make it through a year, but it's just an inefficient way to deal with the problem." Tim and Bill shared this view, noting that large-scale killing is not a good option for reasons of both efficacy and its potential to incite public opposition. As Tim noted: "You have to consider the sociological—because that's the setting in which all of this is going to take place—and if the methods aren't acceptable to enough people, they aren't going to happen."

The techno-tort is a product of, a direct response to, this complexity. With this context in mind, Tim and Bill have set themselves the task of finding better modes of intervening in raven-tortoise relationships. Like Roy and so many people working on this issue that I spoke with in the Mojave, Tim and Bill emphasized that the only long-term solution is to address the root causes of both the increased number of ravens and the wider decline of tortoises (many of which are the same causes, such as development and habitat disturbance). As Bill explained, this involves "getting at the human behaviors, the human activities." Of course, 3D-printed tortoise shells, no matter how well equipped, aren't going to do this, but they might just "stop the hemorrhaging," as Tim put it. He added:

"Until we can alter [these human behaviors] we need to buy time, and we can buy time by altering raven behavior. That's the hope."

Tim and Bill bring to their work decades of research on tortoises and corvids. From 1978 until just a few years ago, Tim spent much of his adult life observing and monitoring tortoises in the Mojave and Sonoran deserts. Over this period he witnessed the decline of the species firsthand until, in 2011, "the only live juvenile tortoise . . . that was seen on the [large-scale monitoring] project, and it was seen by a volunteer, was in the beak of a raven; she saw it flailing its legs as it got carried away." It was around this time that Tim decided to stop simply counting tortoises, and founded Hardshell Labs in an effort to create some new, practical means for intervention. It was through this work that Tim and Bill began their collaboration. Bill has likewise spent decades in the desert, in his case primarily studying ravens as an employee of the Federal Bureau of Land Management (BLM), the U.S. Geological Survey (USGS), and then as a private consultant. His research has explored raven distribution and abundance, their foraging ecology, and their behavior. Bill is one of the world's leading experts on *Corvus corax*, which is no small achievement given that it is "one of the most widespread naturally occurring birds in the world" (Boarman and Heinrich 1999, 1). Having spent most of his career working in the region, Bill has also increasingly been drawn into tortoise research, in particular research on raven-tortoise interactions. It was while planning a monitoring study on the impacts of an urban area on tortoises that Bill came across the techno-torts. He reached out to Tim; "we talked, and the rest is history," Bill explained to me.

Tim and Bill's efforts sit within a larger context in which, in recent years in particular, the pressure to "do something" about ravens in the desert has begun to grow considerably as a result of both conservationists' concerns for the future of the desert tortoise and the concerns of a wider set of interests over a range of associated impacts on industry, agriculture, and urban development. Groups like the Coalition for a Balanced Environment have emerged, advocating for "enhanced raven management" to restore "balance to the California Desert ecosystem."[8] At the start of November 2016, I sat in on a Raven Workshop organized by the USFWS against the background of this growing pressure. The workshop was primarily attended by scientists and managers from various state and federal agencies and universities, but representatives of a range of community groups, industry, and the military (who manage large areas of land in the desert) were also present. Over the course of the day, it became increasingly clear that for many of the people in the room "doing something" about ravens meant large-scale killing. While many people spoke out against such an approach—again, referencing concerns about efficacy and a lack of public support—a range of other speakers and audience members called for the raven to be removed from the Migratory Birds Treaty Act, hoping that this would allow states to get on with the business of killing without federal intervention (a particularly popular idea in states like Nevada and Idaho, which have long killed ravens in much higher numbers than in California for both conservation and agricultural reasons). Other speakers, like one employee of a federal agency, expressed a desire to just "wipe them out" using poisoned eggs. From her perspective, it is only with this kind of "wiping clean of the slate" that parallel efforts to reduce anthropogenic food and water subsidies might hope to be effective. Following this comment a member of the audience noted his agreement, adding that we need to stop access to dumpsters for "rodents, whether flying or otherwise." Among these people there seemed to be a sense that killing ravens, if it can just be done in great enough numbers, will have the desired effect.

The ideas and motivations underpinning this push for large-scale killing are complex. For many of these people this proposal seems to be driven by a genuine concern for the future of the desert tortoise and the hope that reducing the raven population in this way might at least buy time to address broader issues. But there are definitely other forces at work here, too. For others, keeping the focus on ravens helps keep it off precisely these larger social and economic changes—both to reduce the raven population and to help the tortoises in other ways—including reductions in urban development and tighter controls on waste, agriculture, industry, and off-road vehicle use. If the ravens can be presented as the key problem and their numbers can be "reduced," there's no need for these other, much more unpopular measures. As Bill put it: many of these people would like to believe that "if we go after the ravens we don't have to worry about stopping developments because we've taken care of the raven problem." Likewise, Roy noted that many local counties and businesses "don't like the regulatory burdens that come with desert tortoises being listed [under the ESA], so they want to increase the tortoise population by killing ravens.

It's all about killing ravens and not about how we're contributing to the problem." Alongside conservation concerns, ravens seem to be having an increasingly significant impact on agriculture by damaging crops and irrigation infrastructure (to get at the water, or perhaps sometimes just for fun?). In the face of developers, agriculturalists, conservationists, off-road vehicle users, and others, it is perhaps just simpler for many politicians and government managers to keep the focus on the ravens. Unlike all these other groups, Bill noted, ravens lack an organized constituency to lobby on their behalf: "It's really easy to blame all the problems on ravens, [but] ravens are not the primary problem for tortoises." Yet, for all this complexity, they are still *a* significant problem. As such, in the coming years it seems increasingly likely that something is going to be done about growing raven populations in the Mojave. The question is what form this "something" will take.

LASERS, DRONES, AND VIDEOGAMES

We sat in the car in an effort to remain as inconspicuous as possible. From the passenger seat, Tim steadied the laser gun, aimed, and fired at a small group of ravens perched on a power pole a few hundred meters away. Through my camera's zoom lens I saw the bright green dot of the laser appear and move around the upper body of one of the birds. For a moment nothing happened. Then, a few seconds later, the bird began to shuffle uncomfortably, shaking her wings as if simultaneously trying to move away from and dislodge this strange, ethereal parasite. Tim explained: "When they first encounter it [the laser], it takes a little bit longer. Then they get sensitized to it." I asked him: "Do they feel it or just see it?" "I think it's painful in their eyes," he said. "Like a bright light," Bill added. "It doesn't heat their eye up or anything . . . At this distance they're not feeling a burn," Tim said. A second later, the raven was up in the air, followed closely by her companions, seemingly having had enough of this unfamiliar intruder.

We were on a dirt road behind Tim's house in Joshua Tree, looking down over the town and back to the highway. Tim had noticed this group of ten or twelve ravens a few days earlier and then seen them visit the same backyard each day. He speculated that they might be eating pet food or even being deliberately fed. But today they were also being targeted. Tim

FIGURE 4.5. Ravens scattering as Tim demonstrates the use of the laser deterrent. *Source*: Photo by author.

took aim at another group in a large palm tree. As before, they stirred and then moments later took to the air. Most of them circled around for a minute or so and then landed again in the palm or on a nearby power pole. Tim persisted with the laser, taking aim at each of them as they settled. Where at first it had sometimes taken twenty seconds or longer to get them to stir, the effect of the laser gradually became almost instantaneous. The ravens seemed to be becoming more and more worried by it.

After about three minutes of this, with the ravens generally unsettled, Tim put the laser down and suggested that we might try out another of their gadgets. We drove a little further along the road. As we did, Bill commented that he thought it would be fine for us to get out of the car: "I don't think they'll scatter . . . especially since we're in an area with people." Tim added: "They're used to people moving around, and I don't think they think the car is the source of the laser." As the two biologists conferred on the capacity of these ravens potentially to recognize us in this way, I thought of the research by John Marzluff and colleagues that demonstrated that corvids, in this case American crows (*Corvus brachyrhynchos*), can not only recognize individual humans but are able to form negative associations about them/us and to remember and pass this information on to others (Marzluff et al. 2010). This research showed that crows use facial recognition to make these distinctions among humans but also that it's not just faces they're attentive to: the crows they studied were also paying attention to their car and to particular dogs, and again they were able to distinguish between similar entities (Marzluff and Angell 2005). Tim and Bill have firsthand experience of this corvid capacity for recognition, so much so that in some cases it has become difficult for them to test the laser and other devices. As soon as they show up in their car at an industrial composting facility or other popular raven hangout, thousands of birds take to the air, presumably knowing what's coming.

As we got out of the car, I commented to them on the potential difficulty of this situation: "I guess with most birds you don't really have to think about them recognizing you or your car." "Yeah," Bill laughed. Tim added: "But that's part of the pleasure of it." "The *pleasure*?" I asked. "It just makes the whole thing more *interesting*," he replied. "You've got an animal with the capacity to recognize you as an individual, but you don't have the capacity to recognize *it* as an individual, short of marking it." In this particular case, however, the ravens had no history to draw on; they didn't recognize us or the car, and so they slowly settled back in after the excitement of the laser, only to be greeted moments later by a drone.

Tim placed the drone on the ground in front of him. It was a squarish, orange device about a foot in diameter with four large white propellers that covered most of its upper surface. As he stepped back from it he called to us to watch the two ravens on the nearest power pole. With the remote control in hand, he started up the propellers, and as the drone lifted off, so did the birds. "Oh, they're already gone," he laughed. "And there's two more going. As you probably know, ravens have this characteristic called 'neophobia,' and this is definitely new for them." Tim took the drone further up into the air, closer to the power pole and the palm tree where most of the remaining ravens were perched. In each case, as it approached, the birds took to the air. He didn't need to get close to them at all. As he demonstrated the drone, we spoke. Tim told me that in addition to this general hazing, he imagined that drones might even be fitted with lasers and perhaps even a methyl anthranilate sprayer, anything to keep things interesting, to keep the birds guessing. After only a couple of minutes all of the birds in the area seemed to get the message: things weren't going to quiet down any time soon. About fifteen ravens began to kettle, riding a thermal in a circular motion high up into the air above us, taking advantage of this lift to move on more easily to another, hopefully less problematic, part of the desert.

The basic principle behind the use of both lasers and drones is the same. The neophobia—fear in the presence of new or unfamiliar objects—that Tim mentioned as he launched the drone is crucial.[9] In a later conversation he elaborated on this point: "I don't know what they're thinking when they see this stuff, but they don't like it. They're conservative and really cautious, and that's our huge advantage. . . . They're utterly practical animals." On encountering a laser or a drone, ravens tend to move on quite quickly, with the odd exception who tries to hold his or her ground. For a few weeks just before my visit, Tim and Bill had been out at a local pistachio farm trialing the laser and monitoring its efficacy. They started by counting undisturbed ravens in the orchard for several days. Then for the next three weeks, from the cover of a blind in an elevated tower, they fired the laser at any ravens who arrived to scare them off. Not only did they manage to keep the orchard clear of ravens for this period, but they found that for the next few weeks, even without the laser, almost all of the ravens kept clear. While the odd raven seemed more resistant to the laser or returned more quickly, refusing to give up on the bounty, the fact that ravens are also profoundly social—and therefore less willing to eat alone—meant that these individuals all ended up abandoning the site anyway.

It seems, therefore, that both lasers and drones can be used to drive or keep ravens out of an area. But what might this mean for the countless ravens, not to mention the declining tortoises, of the vast Mojave Desert? Back in Tim's kitchen, over a large bowl of pistachios, I asked him and Bill about this: were their projects really scalable in the way that they would seem to need to be? The simple answer was that they weren't sure, but they were optimistic. This optimism, I discovered, arose out of a thoroughly pragmatic reframing of what meaningful action on this issue might look like. Grounded in decades of work in this place, on this issue, both men seemed to have arrived at the same realization: there would be no singular solution, neither a management action that could be carried out once and for all nor a single approach that could simply be repeated in perpetuity. Any "solutions" to the "raven problem" were going to be much more partial and piecemeal, cobbled together out of minor successes that might just buy enough time to address some of the many other vital areas of tortoise threat that require more significant social and political change.

Far from excluding ravens from the desert altogether or preventing all tortoise predation, the intention behind the lasers and drones is that of intervening at key sites to keep ravens out of areas of abundant, vulnerable, tortoise habitat and deny ravens access to important resources. For example, they envisage that lasers and drones might be used alongside current "head-starting" and translocation programs to create large "no-fly zones" that would function something like marine protected areas, but in this case limiting raven rather than human activity. Current head-starting facilities bring free-living female tortoises into desert enclosures to lay their eggs. When they hatch, the young are protected by the netted enclosure for their most vulnerable years. As they get larger the juvenile tortoises are released into the wider desert. With lasers, tortoises might be released earlier or provided with some additional protection for a bit longer. Similarly, with regard to resource denial, ravens might be kept out of rubbish dumps, composting facilities, sewage-treatment facilities, farms, and other highly localized sites that provide year-round access to abundant food and water, or perhaps they could even be denied access to the power poles that in vast areas of the desert are the only available nesting spots. Some measures have already been made to limit access to these sites, but lasers and drones might supplement and enhance these approaches. If drone and laser technologies were deployed in this way at a reasonable scale, they might even lower the raven carrying capacity of the desert. As Bill put it: "If we're denying them key resources, they're not going to be able to raise as many young."

One immediate question that follows is: who will operate all of these lasers and drones? Here too, the tech edge of Hardshell Labs is clear. Bill and Tim are banking on, and indeed already hard at work generating, technological solutions, in this case in the form of artificial intelligence (AI) driven by pattern-recognition software and, perhaps even more ambitiously, remote, crowd-sourced labor that will be harnessed through videogame interfaces. Alongside providing this labor, Tim hopes to produce games that can be genuinely informative and perhaps even transformational, prompting players to care in new ways for threatened species and their ecosystems. Some games might allow gamers to operate a laser tower or a remote-controlled car or drone that keeps ravens out of designated areas, while other, slower-paced games will involve weeding the desert or observing tortoises and scoring points for "natural history acuity."[10]

It is worth noting in this regard that while Tim and Bill have their sights firmly fixed on conservation, they're well aware that there are a range of other applications for their endeavors. If drones and lasers can be used cheaply and easily to keep ravens (and perhaps also a range of other birds) out of designated areas, then this is something that orchardists and agriculturalists the world over would likely be interested in: a twenty-first-century scarecrow that actually works (Lorimer 2013). In some parts of the world, like the Mojave, this might have positive conservation outcomes; in other places the situation will likely be very different.[11] At this stage, with all of these technologies still in the planning and development stages, it is too early to say how these different possibilities will take shape. Tim and Bill are far from naïve about what is at stake here. They are, however, cautiously optimistic. As Tim put it: "Any time we open the technology box it's a mixed blessing . . . but the technology is going to develop, and my attitude . . . is how about we get in front of it and try to steer it in beneficial ways to the degree that we can."[12]

RECOGNITION AND DIPLOMACY

In a range of different ways, questions of recognition are at the heart of these proposed interventions into human-raven-tortoise interactions in the Mojave. How do we appear to one another? Who attaches what meanings and possibilities to another, or fails to? In short, we are drawn here into the messy question of who we understand ourselves to inhabit the world with, and with what consequences, and for whom. In its various uses in the English language, recognition describes acts of intellectual apprehension (to recognize a fact); of visual, audible, or some other sensate identification (to recognize a friend or a sound); and of acknowledgment and respect (to receive adequate recognition, perhaps in terms of treatment or rights).[13] In these Mojave encounters, there are many different modes of recognition at work and at stake, cutting across these loose categories; there are also many different literatures on recognition that might inform our understanding of these processes. In philosophy and politics, as Paddy McQueen notes, discussions have tended to focus on the last of these three forms of recognition, namely acknowledgment and respect (McQueen n.d.), although the borders are never as clear as we might think (Margalit 2001, 128–29). In the biological literature, recognition—not always referred to in precisely this way—gets mixed up with questions of "other minds," notions of self and other, and even with the possibility of empathy (de Waal 2008, Bugnyar 2007). Influenced and inspired by all of these literatures, my thinking here plays back and forth between different registers to tease out some of what is at stake in these particular ways of paying attention to and being with others.

Of critical importance to this Mojave story is the way in which these Hardshell Labs proposals are grounded in the fundamental recognition of ravens as lively, intelligent beings capable of particular modes of learning and response, even while the parameters of these specific capacities remain shifting and uncertain and, as such, needy of ongoing experimentation and attention. While shooting or poisoning ravens demands very little of them as intelligent beings—in fact, animal intelligence is often simply viewed as an obstacle to be overcome in this context, for example, ravens identifying and avoiding a shooter—things are markedly different within the context of these proposals. Tim and Bill need to concern themselves with questions like: do the ravens really (mis)recognize a 3D-printed shell as a tortoise? The whole efficacy of this proposal rests on them doing so. If the techno-tort is easily distinguishable from the "real" thing, then even if Tim and Bill can sort out all the mechanics, ravens won't be learning the right lessons from exploding shells—they won't be learning to avoid tortoises but rather just techno-torts (a possibility Bill is currently testing experimentally). In taking up this approach, they need to think about where this tortoise-eating behavior comes from: for example, is it a product of social learning or individual innovation? Likely a bit of both, Bill suspects, but there is more to be learned here. It may even be possible to, in Tim's words, utilize the techno-tort as a kind of "social hack" in which raven communication and learning networks help spread the word in some way. It is with this in mind that he has asked: "can an 'informed' raven be more valuable than a dead one?"[14]

Meanwhile, the laser and drone projects raise important questions about whether ravens will eventually become habituated to these technologies. This is something Tim and Bill are also actively exploring. As Bill put it, with reference to the laser: "will they at some point realize that it's not doing anything and learn to ignore it?" For his part, Tim thinks that the laser does cause them mild irritation, perhaps even a little pain, like a camera flash going off in their eyes. But will they one day "start to grit it out" instead of taking flight? Or, conversely, might they be becoming hypersensitive? Tim thinks he is witnessing this already as ravens seem to come to understand what the green light means and to vacate an area more readily. Here, raven capacities of memory, learning, and becoming sensitive in different ways to new stimuli are all vital—even if somewhat opaque—components of the project. Likewise, as we have seen, raven sociality and neophobia are crucial: the former meaning that far from having to "haze" every individual bird, once a few individuals become uncomfortable and take to the air, the rest will follow, while the latter ensures that ravens faced with a potential threat like a drone tend not to resist bravely or mount an offensive but instead just leave.

In short, the general philosophy that underlies the approach being developed here is one centered on a respectful awareness of both the challenges and the opportunities created by these birds' intelligent, adaptive ways. Attention to the nuances of raven behavior and biology inform, they *make possible*, these approaches. Responding to a "thick" sense of ravens—of who they are, why they act in the world as they do, and what they are capable of—Tim pointed out: "This will be a challenge for us too. As we're doing our thing, they're [the ravens] going to adapt to it. If the motivation to exploit a resource is intense enough they'll try their best to work around whatever we come up with. But that also makes it kind of fun. It's a fun process. You've got this other brain involved, an alien brain. It's a chess match." It is for this reason that Tim and Bill view each of their individual projects as simply one "tool" in an ever-evolving set of resources. If ravens learn to ignore lasers or to recognize techno-torts for what they are, they will need another option to wheel out in its place or an upgrade or modification to shift the balance again.

Ravens emerge here as complex *subjects* of a world. As a great deal of scholarship has shown, the past several hundred years of Western thought have been dominated by notions of animals as "objects," perhaps playing out instinctive or genetic programs, "sleepwalking through life" (Crist 1999, Chrulew 2017a, Lestel 2014b, Buchanan 2008, Despret 2008b, Steiner 2005). These objectifying approaches have often strongly influenced wildlife management and biodiversity conservation, sharply constraining the kinds of approaches that might be taken—a situation that is being increasingly challenged, however, with calls to better integrate behavioral biology into these areas (Berger-Tal et al. 2011). This is the "other" culture that is slowly finding its way into dominant efforts to understand and live with wildlife in many parts of the world; the more-than-human cultures, alongside the diverse human cultures, previously excised or ignored. In this context, recognizing ravens and at least some other animals as subjects (of some kind) is part of what Dominique Lestel (2014b, 114) has called "the true scientific revolution in the animal science of the past twenty years: *the human being is no longer the sole subject in the universe.*"

In drawing attention to ravens as subjects in these ways, I want to create some distance from notions of the "actant" or "agent" in the minimal sense of these terms: that is, making a difference, modifying a state of affairs, like billiard balls knocking into each other, exerting an effect. There is a tendency in some recent work in actor-network theory, new materialisms, and related fields to focus on as thin a version of agency as possible so that it might be recognized in more and more parts of the world (Bennet 2010, viii; Latour 2004, 237). While there is tremendous value to such an approach in certain contexts, we also require "thicker" notions of subjectivity to make sense of the diversity of beings that share and constitute our worlds. Importantly, ravens are not just agents in this minimal sense; they are also beings with their own understandings, their own modes of paying attention, of learning, remembering, becoming sensitive, and adapting understandings and behaviors. I want to hold onto a notion of subjectivity as something distinctive in this way. *To be a subject is to inhabit an experiential world characterized by embodied experiences, impressions, feelings, understandings, and beliefs.* As Peter Godfrey-Smith (2016) has succinctly put it: "For some animals, there's something it feels like to be such an animal. There is a self, of some kind, that experiences what goes on."[15]

My goal here is not to demarcate *a* way of being but rather to gesture toward a general mode of being that is undoubtedly taken up and experienced in myriad ways by diverse living beings. How and in what ways different species, and indeed different individuals, "do" subjectivity remains a question for ongoing empirical study and reflection. In this context, subjects are not unitary and self-forming; nor are they some sort of "mental" element that can be divorced from the material. They are "trans-corporeal" and "exposed" (Alaimo 2016), woven into and through one another. This is, as Rosi Braidotti (2013) outlines, a notion of subjectivity "as an assemblage that includes non-human agents," one in which the subject is understood as a "transversal entity encompassing the human, our genetic neighbours the animals and the earth as a whole." Of course, subjects are also bounded in important ways, too. They/we are ongoing, multiplicitous, nested, and entangled becomings, modes of being that emerge and continually reemerge inside long histories of biosocial intra-action and inheritance.[16]

In recognizing ravens as particular kinds of intelligent, adaptive, sociable subjects, these Hardshell Labs proposals open up the possibility for less lethal modes of engagement; more specifically, they enable what Vinciane Despret has called a "diplomatic" response. Thinking with her friend and collaborator Isabelle Stengers, and in conversation with Baptiste Morizot's (2016) recent work, Despret (2016a) explores the possibility for human-wildlife interactions grounded in "influence" rather than "control." In place of efforts simply to dominate another, perhaps lethally, to exploit their weaknesses, diplomacy seeks to work with another's strengths; it seeks to generate new possibilities rather than to impose one party's predetermined order. As Stengers puts it: "diplomacy presupposes a peace to be invented, not the weak part bowing down in front of the strongest part" (Zournarzi 2002, 271); it involves the recognition that "nobody can speak in the name of this situation. . . . It implies, for each involved party, different risks and different challenges" (Stengers 2005, 193). In a multispecies context, diplomatic negotiations must take complex forms, but Despret and Morizot are clear that shared "language" and forms of "rationality" are not necessarily required. Instead, Morizot notes that "it is sufficient for another to be, for example, social, territorial, smart. Is s/he required to be willing to negotiate? No, we just need the other to be able to receive messages."[17] And so, we need to be able to "formulate these messages" in a way that *proposes* something to the other, to interpret their response (alternative proposal), and then perhaps to propose ourselves again (Despret 2016a).[18]

One of the key examples Despret (2016a) borrows from Morizot to illustrate this possibility is the use of masks to limit tiger attacks on people in the Sundarbans in India. Knowing that tigers often attack when their prey is not watching them, people wear masks on the back of their heads. "The ethological ruse," Despret explains, with reference to Morizot, "is to deprive the tiger of the stimulus that provokes the attack." Morizot notes: "Neither the gun, nor the trap, but the mask became the freak weapon of this zoocephalic diplomat to protect himself from the tigers . . . with a minimum of strength, and a maximum of pragmatic, ecological and political effects." The mask is referred to here as a "ruse" and a "freak weapon," but it is also a diplomatic communiqué. As Despret notes, formulating diplomatic messages "involves what I would call the art of ruses and tricks" (13). From this perspective, negotiations are not at all about transparent, rational deliberations in which each party clearly lays out their views and desires. Instead of this caricature of an idealized space, we are offered a much messier, more compromised account of what it means to engage in diplomatic negotiations.

The tools being developed by Hardshell Labs are precisely this kind of diplomatic proposal. What is essential here is the way in which these tools respond to particular raven characteristics, working back and forth in an ongoing process of proposal, adaptation, and change, to produce an outcome that neither side can quite anticipate. Techno-torts send a clear message to ravens: tortoises are not edible. In so doing, they propose a new set of relationships, a new way of getting on together, which ravens will test and perhaps accept. Of course, if they do not accept, Tim and Bill will likely attempt to rephrase the proposal: perhaps a new triggering mechanism or nontoxic deterrent will express the message more clearly and forcefully. Likewise, lasers and drones might announce to ravens that these anthropogenic resources, or these particular areas, are now off-limits. Again, the ravens may or may not accept these proposals, and they will need to be rethought or reformulated. So it will go, back and forth. There will be no final, perfect peace, but the work of diplomacy may just prevent the kicking off of efforts to dominate ravens lethally, to establish forcefully the kind of "peace" that Stengers has referred to as a "police operation" (Zournarzi 2002, 271).

There are many possibilities here. Diplomacy is the subtle art of working with-and-against another to produce mutually livable worlds. As such, it is a practice that demands more of both ravens and people than killing does: "Poisoning is easy but nurturing is a craft" (Stengers 2008, 38). As previously noted, beyond simple questions of dead or alive, of population numbers and their impacts, these diplomatic projects demand that attention be paid to ravens *as subjects*, to their processes of learning and communication, to shifting understandings and sensitivities, to modes of becoming-otherwise. Recognizing another in this way involves the

FIGURE 4.6. Desert terrain, Mojave Desert, California. *Source*: Photo by author.

negotiation of sameness and difference, self and other, as Val Plumwood (1993) emphasized in her work on the topic.[19] For Plumwood, recognition is about the appreciation that we are encountering another being who is a locus of thought and action, an entity "alike in being a centre of needs and striving," but at the same time, it is a question of difference: "recognition of the other as a limit on the self and as an independent centre of resistance and opacity" (157).

To refuse, or fail, to encounter others in this way is to inhabit a world of objects, to adopt a stance of domination and instrumentalism in which others are only relevant, and indeed legible, in so far as they are of use.[20] Returning to Despret, Stengers, and Morizot, this is a world in which diplomacy is simply impossible: the other is neither capable of responding to nor worthy of diplomatic overtures. To enter into diplomatic relations simultaneously *requires* this kind of recognition of another while also being itself a further *act* of recognition. In short, in taking up their particular diplomatic proposals, Tim and Bill both recognize and insist on the existence of *another raven*: one that can and should be engaged as a wily, creative, adaptive being.[21]

There is, of course, a danger here that the language of diplomacy might be used to cover over very significant inequalities in power and the capacity to resist and effectively propose alternatives. As Matthew Chrulew (2017a) has argued, the recognition of animal subjectivity, agency, and mindfulness is not always an entirely positive thing—an assumption that is often made in animal studies and environmental humanities literatures. Instead, biological understandings of animal behavior and the specific dimensions of animal subjectivities have often been leveraged to optimize and intensify regimes of biopolitical management in the zoo, the factory farm, conservation projects, and beyond (Chrulew 2017b, 2017a, 2011a). This kind of biopolitical management substitutes the lethal intervention of the sovereign for the ongoing disciplining, reeducation, and so redoing of the ways of life of individuals and ultimately whole populations. In making this point, Chrulew reminds us that the line between diplomacy and biopolitics (with its own forms of domination and control) is a blurry one. As such, what is needed is critical attention toward both the objectification of others and the particular regimes of subjectification that they are required to live within.[22]

From this perspective, the Hardshell Labs proposals might readily be thought of as exemplars of a new form of biopolitical management. Diplomacy and biopolitics are not mutually exclusive possibilities. Distinctions here are complex and shifting. To explore interactions like those in the Mojave as sites of diplomacy is not to assume that they are in some way benign or free of extreme power asymmetries and indeed violences (when was diplomacy ever this way?). Rather, it is to draw attention to these sites of engagement in a particular way, namely, in a way that takes seriously the subjectivities, the modes of understanding and worlding, of *all* of the parties involved, as well as their genuine, even if always limited and unequal, capacities to respond in their own ways (Palmer 2001). The recognition of another's subjectivity is a necessary but not a sufficient condition for diplomacy. Such diplomacy is always fraught: it is the work of being and becoming with other subjects, *for good and ill*. In contrast to such an approach, the effort to dominate completely or exterminate another must be understood as the denial or negation of their subjectivity, of their existence as an "other" who is both capable and worthy of engagement in a fuller way.

Of course, in many cases diplomacy is simply the result of the fact that the other *cannot* be eradicated or controlled in an absolute manner. As such, it seems that while diplomacy might require the recognition of certain capabilities and modes of being in the other, it need not require the recognition of the other as making a claim on me, as being in some sense *worthy* of such a response. This distinction matters. While there is great diplomatic potential in the projects being developed by Hardshell Labs, does this mean that they are grounded in the recognition of ravens as beings who *ought* not to be managed lethally? If, somehow, killing ravens en masse were to become both an effective means of limiting their impacts on desert tortoises and a palatable operation for the wider public, would Tim and Bill still opt to pursue the intricacies of lasers and exploding tortoise shells?

There is no simple answer to this question, and in hindsight I'm glad that I didn't think to ask it at the time: I don't think it can be answered well in the abstract. At various points in our conversations Tim and Bill did both state that they would rather not kill ravens: they preferred the possibilities of diplomacy. Both also made clear, however, that they were not opposed to killing for conservation in any general or categorical way. But, while many other conservationists and wildlife managers have continued to push for large-scale poisoning, Tim and Bill are doing something different. As Bill put it: "it's not that I'm looking for any and all solutions to avoid having to kill ravens, but it feels much better to not kill them." Tim added: "This seems much more interesting and attractive than poisoning massive numbers of highly intelligent creatures, or blasting them, or whatever"—something which, he noted, would likely involve "having to kill tens of thousands of ravens, and having to do it year after year after year."

There are no categorical imperatives here, no room for an ethics of purity (Shotwell 2016). Instead of absolute positions—for or against killing—we are faced with a much more contextual desire not to kill unless it becomes unavoidable, perhaps unless some other competing obligation requires it. As our discussion about the ethics of killing was drawing to a close, Tim made an interesting observation: "I don't think the lethal-control advocates *like* killing animals either," he said. "They just don't see another way." I'm sure he is right about this. While some people undoubtedly do *like* to kill and harm animals, there's no reason to believe that these particular lethal-control advocates are among them. But herein lies the really challenging ethical domain. For those who don't like to kill, the question is precisely how one responds in the face of this possibility. First and foremost, this is an issue of recognition: who will count as a meaningful *other*, a being worthy of consideration, as opposed to, say, a "flying rodent"? Beyond this, however, it is a question of translating that recognition into the concrete possibilities of diplomacy: How creative, imaginative, and dedicated will we be about seeking and indeed crafting "another way"? How much are we willing to put into a diplomatic solution before moving on to projects of domination, control, and killing? It is in these imaginative spaces and practices of alternative-making that a core part of the work of ethics as worlding takes place. Making flourishing worlds with others requires an ethics of creativity, of refusing preformulated choices and understandings, an ethics that asks again and again: *what else is possible?*

In this context, thinking through the lens of recognition is important precisely because it opens up new possibilities for creative engagement with others. This is about, wherever possible, recognizing others as other subjects, same and different, capable and worthy of diplomatic overtures. Such an approach does not guarantee that there will not be violence and killing. It does not even guarantee that the results will be better. These are spaces of "anthropo-zoo-genesis," in Despret's (2004) terms: of becoming human-with-raven and raven-with-human, as we always are, but *anew*. However, in entering into these relationships diplomatically—in a way that actively seeks to recognize more of what another is capable of, of what matters to another, of another's intelligent, responsive, creative potential—we open up possibilities for more attuned and mutually enriching ways of life.

LOOKING BACK

Everywhere I traveled in the desert I looked for ravens. As I drove each day I gazed upward, as I walked in towns and national parks I kept an eye and an ear open, as I sat and read or ate I watched out. I also sought out dumpsters and rubbish tips, places more likely to provide an opportunity to encounter ravens up close. Toward the end of my time in the area, in search of a last encounter, I headed back to Joshua Tree National Park, to a big picnic area I had seen earlier. The parking lot was full when I arrived, with a couple of large tourist buses and thirty or forty cars. In several spots groups of people were eating their lunch at benches. I sat and waited. Sure enough, about ten minutes later, two ravens arrived. They moved around largely unnoticed. They watched people eating. They perched and occasionally cawed loudly. At one point I moved closer to them for a better look. As I got within about ten meters one of the birds fixed its gaze on me. I stopped and looked away, hoping to defuse the situation. The raven in turn looked away. But as I turned back about ten seconds later, so did the raven, this time taking to the air in response. This pattern of interaction repeated several times over the next half hour, as it has so many other times in my years of corvid watching. The ravens watched and moved about among the people, sometimes in relatively close proximity, but as soon as they became aware of being watched, by me or someone else, they took off. A casual glance was fine, but a directed look for even a moment caught their attention (see figure 4.1).

This sensitivity seems to be a general corvid characteristic. As the biologist Thomas Bugnyar pointed out to me in an interview: "Corvids are very sensitive to where other animals are looking, and they don't like to be the focus of attention." He speculated that this might have something to do with their scavenging lifestyle, in which they "eat food which is also of interest to other animals, so this makes them wary." In experimental work, corvids have been shown to be good at reading "gaze cues" from both other birds and humans. For example, research on jackdaws (*Corvus monedula*) has found that birds were less likely to approach food if an unfamiliar person was looking at it (von Bayern and Emery 2009). Regardless of the person's facial or bodily orientation, these jackdaws paid attention to people's eyes. This is a skill that, as the researchers noted, may have evolved through its important role in interactions with conspecifics (other jackdaws). Here, corvids are helped in their interactions with people by the fact that, unlike many other animals, both of us have eyes with a white/light-colored sclera and a dark iris or pupil, which "makes . . . eye orientation discernible" (von Bayern and Emery 2009, 602).

In paying attention to raven eyes, to ravens looking, we are reminded that they look back at us. It is not just humans with the capacity to pay attention to others. Perhaps it is not just humans who are able to themselves recognize others as *other subjects*, able to, as Plumwood (1993, 157) put it, enter into relationships in which they are "aware of one another as significant others." To recognize others as subjects in this way is, in many cases at least, to cultivate an awareness of others' own modes of recognition. This is likely something that species are engaged in in distinctive ways. In the case of corvids, however, we are fortunate that, yet again, this is a possibility that has been experimentally explored over the past several decades.

In 2003, in an aviary in Vermont, on the other side of the United States, Thomas Bugnyar and Bernd Heinrich offered captive common ravens (the same species found in the Mojave) some tasty morsels and various opportunities to cache them away from the prying eyes of other ravens and humans. Beyond simply reading their gazes, this experiment aimed to determine whether these birds can make sense of others as knowing beings with their own understandings and agendas (Bugnyar and Heinrich 2005). In this and other similar studies, what is at issue—in the parlance of biology—is the existence of a "theory of mind" (ToM). Long assumed to be the sole province of "the human" (Haraway 2008, 235–37), a range of animal species are now thought—by at least some behavioral biologists—to possess something like a ToM. To have a ToM is to be the kind of creature that understands that "the behavior of animate beings is guided by perceptions, attentions, intentions, and beliefs" (Bugnyar 2007, 15), in short, to have a sense of oneself and at least some others as loci of thought and decision making, with their own distinctive projects and desires.

Bugnyar and Heinrich's experiments showed that ravens who were caching food items away to recover later distinguished between both other ravens and humans who knew where particular caches were located and those who didn't, acting preemptively to ward off the theft of "knowers" while ignoring the movements of "nonknowers."[23] As the researchers put it: "Pilfer and antipilfer tactics suggest that ravens judge their competitors on the basis of what they remember them paying attention to. They then attribute to the competitors the capacity of knowing, and they integrate that knowledge together with dominance status into strategic decisions for making and retrieving caches" (Heinrich and Bugnyar 2007).

Based on these and subsequent experiments, Bugnyar concludes that ravens clearly have "some mental representations about others," that they "can encode information that is 'unique' to the perspective" of another and utilize it in decision making in relation to food caching (Bugnyar

2010, 5; Bugnyar and Heinrich 2006). The precise contours of their knowledge about others, and the contexts in which such knowledge might be applied, remain speculative. Whether this behavior is grounded in what we might think of as a fully formed ToM is unclear, but at the very least it does provide evidence that ravens understand some other living beings as something like "subjects."[24]

In corvids, this capacity is also accompanied by a keen ability to recognize (here in the sense of "identify") and remember distinct individuals. As previously noted, Marzluff's research in this area is key. In studies conducted on the American crow, he and his collaborators established that a bird "quickly and accurately learns to recognize the face of a dangerous person," information that is retained "for at least 2.7 years" (Marzluff et al. 2010, 699). Birds exposed to trapping by humans wearing face-like masks as they worked would accurately identify individuals wearing those masks and loudly scold and mob them. As different people, with different physiques and walking gaits, etc., could wear the "dangerous masks" and receive the same treatment, it seems that the crows were utilizing specific facial features to recognize potential threats. As previously noted, anecdotal evidence indicates that they're making similar discriminations in relation to dogs, cars, and presumably other entities too. While detailed studies on the identification of individual human faces have not been conducted on/with common ravens, there is every reason to think that they too possess this ability: in fact, experiments like those on caching and the field experiences of people like Tim and Bill strongly support this conclusion.

From these kinds of experimental observations we might conclude that corvids inhabit a "peopled world": a world full of distinct subjects, beings with different histories, understandings, and projects. The term "people" here is anything but a synonym for "humans." As Graham Harvey (2006) notes: "the world is full of persons, only some of whom are human." This does not mean that ravens understand personhood or subjectivity in the ways that humans do (which one?). Rather, to make this point is simply to acknowledge a barely graspable but nonetheless tangible similarity between human and corvid modes of being. This is a similarity that is likely not shared with a wide range of other nonhuman animals, at least as far as we know. So, while as discussed in the previous section, we might recognize other living animals as subjects, this certainly does not mean that *they* all understand themselves, or others, in similar ways. Ravens, however, seem to be among those beings that do.

As subjects of recognition, ravens are capable of *inter*subjective engagements with us and with one another, a possibility that requires "the mutual recognition of sameness and difference in dynamic interplay and the sharing of similar conscious states" (Plumwood 1993, 213). Moving outside the myopic focus on the human that characterizes the work of Hegel and many other key thinkers in this area, there is no reason to believe that we might not be subject(s) to one another in various ways across species divides.[25] Indeed, to encounter another in a moment of intersubjective recognition—to see oneself in and through their knowing gaze—is itself a fundamental challenge to the imagined significance of such divisions. While moments of recognition are always rich with life- and world-forming potential, where genuine forms of intersubjectivity occur, they open up distinctive possibilities for co-becoming. As Donna Haraway (2008, 236–37) notes: "Among beings who recognize one another, who respond to the presence of a significant other, something delicious is at stake." There is what she calls "the taste of copresence," and with it a possibility for the "shared building of other worlds."[26]

Many of the questions posed in this section—those opened up by genuinely intersubjective forms of recognition—likely are not relevant for our engagements with a broad range of other species, engagements characterized by a diversity of embodied beings involved in all manner of modes of relation and detachment (Ginn 2014, Candea 2010, Reinert 2013, van Dooren 2016). But this fact does not limit the importance of paying attention to recognizable forms of intersubjectivity where we do encounter them. Multispecies ethics cannot be a one-size-fits-all proposition; the categories and lenses with the broadest applicability are not necessarily the best or the only relevant ones. Life requires a diversity of modes of ethical encounter, open and responsive to its genuine diversity. The task is not to dissolve difference but to become attentive to its multiplicitous forms.

With this in mind, it perhaps bears noting that all of the experiments on corvid cognition mentioned here focus primarily on visual recognition and engagement. Like (most) humans, corvids and most other birds also rely heavily on their vision. But recognition is not limited to the modality of sight. It is an embodied phenomenon working across diverse sensate and communicative terrains. It can even, as is so powerfully apparent in figure 4.1, take place in acts of looking away, a postural communication of, in this case, politeness that was coupled with hearing (and peripheral vision) to ensure that both parties were honoring the implied agreement.

From the foundational work of Hegel and J. G. Fichte onward, philosophers have emphasized the significance of intersubjective recognition and the way in which our understandings of ourselves are influenced by others' ways of seeing us, including misrecognitions and a lack of recognition. Identities are at stake here, alongside much broader possibilities for flourishing forms of life (Honneth 1995, 169).[27] In this context too, recognition is explicitly *not* a question of preformed subjects entering into relationship with one another. Hegel's thought in this area worked in explicit opposition to the atomistic notions of the self-forming individual that dominated philosophical thought in his day (and still do in many ways). In contrast, his emphasis on recognition was grounded in the understanding that, as Nancy Fraser (2003, 10) has put it, "social relations are prior to individuals and intersubjectivity is prior to subjectivity." In other words, subjects are made and remade in their relatings, reaching into one another, mutually imbricated from the outset.[28]

Paying attention to intersubjectivities in this way is not just, as discussed in the previous sections, about recognizing and responding to another subject's creative and intelligent capacities. In addition, it is about paying attention to the profoundly psychic ways in which our modes of recognition influence others' senses of themselves, of the world, of being and relating, and ultimately their personal and social possibilities. And it is, concomitantly, about asking how we ourselves are remade by the ways in which others come to recognize us. Who might the ravens of the Mojave *become* inside different, perhaps more life-affirming, modes of recognition? Who might we all become together?

Taking up these questions is a necessarily speculative endeavor, but it is perhaps especially so in the context of multispecies communities and other-than-human subjectivities. Yet there are traces of an answer, or at least traces of the kinds of possibilities for good and ill that might arise here. In her research on parrots rescued from abusive homes, Jean Langford (2017, 86) paints a tragic picture of birds who seem to have internalized such a diminished sense of themselves that they engage in ongoing self-harm, both physical and verbal. Some birds pull out their own feathers and stab themselves, others berate themselves in terms they must have heard again and again: "I hate this stupid bird." Far from being empty sounds devoid of meaning, Langford argues convincingly that we might at least hold open room for an understanding in which such an utterance "articulates a surge of feeling that specifically echoes a past situation of hostility." In a related vein, other studies have found that elephants who have survived the slaughter of their herd are "displaying symptoms associated with human PTSD [post-traumatic stress disorder]—abnormal startle response, depression, unpredictable asocial behavior, and hyperaggression" (Bradshaw et al. 2005, 807) and may now even be consuming alcohol as part of their response to inhabiting stressful environments (Barua 2014). Likewise, chimpanzees that have been exposed to traumatic experiences seem to display symptoms of depression and PTSD (Ferdowsian et al. 2011). For those animal species with complex social bonds, trauma manifests in ways that cut across the individual and the collective. Processes of "social breakdown" both arise as a result of and further contribute to the traumatic experiences of individuals (Bradshaw et al. 2005).

As is so often the case in this kind of research, we have primarily behavioral signs to go on. But these signs point very strongly to processes of psychic rearrangement, in which new subjectivities and subjective possibilities are taking shape as a result of trauma. In different ways in each of these examples, animals' modes of understanding and of comporting themselves in the world are remade as a result of human practices of recognition. These practices of recognition always take place through the complex entanglement of modes of understanding, valuing, and *treating* others: recognition is a profoundly and simultaneously material, social, and psychic process.

We don't have these kinds of studies of possible trauma in corvids, but we do know that they are responsive to humans in a range of ways. We know that corvids who live in areas where they're likely to be shot give humans a much wider berth than those who don't, and we know that this behavior is a social one, learned through experience and passed on (Knight, Grout, and Temple 1987). We know that corvids are finely attuned to people paying attention to them, to being watched, and that they learn to recognize and avoid us (von Bayern and Emery 2009, Marzluff et al. 2010). We know also that they are beings with complex emotional lives who, in at least some cases, seem to grieve, to console each other, and to cooperate (Fraser and Bugnyar 2010, Heinrich and Marzluff 1995, van

Dooren 2013). With all of this in mind, it does not seem like too much of a stretch to suggest that the psychic lives of corvids—like those of elephants, chimpanzees, and large parrots—may in their own ways be profoundly influenced by human modes of recognition.

This possibility opens up an unsettling question and with it new kinds of responsibilities: how might particular modes of recognition and their concrete practices of engagement remake ravens' psychic and social possibilities? What kinds of flourishing, but also of trauma, are made possible here? Taking raven subjectivity seriously in this way offers a different, potentially more critical lens on Tim and Bill's proposals. In our discussion, as they elaborated on a string of possible technologies and interventions, it became increasingly clear that what is being imagined here is, in an important sense, an ongoing experience of confusion, uncertainty, and fear for ravens. As Tim explained this overall vision: "My sense is that you have to create this haunted landscape for them, where they cannot relax. They're on edge, they never know what. . . . That's another reason to have all these tools. Don't depend on one, but have a mix [to the point where they're thinking]: 'boy, that valley over there is just weird. Every time I go over there something comes up and I never know what it's going to be.'"

While in some places ravens exposed to lasers and drones might simply move on to other environs—avoiding the "valley over there"—if all goes to plan some birds may well be left with little choice, especially as resource availability diminishes. As this happens, less-dominant ravens may be pushed into this new form of "marginal habitat." What will the psychic toll of inhabiting, or even frequently encountering, these places be? What kinds of raven subjectivities and socialities will be possible in such a place?

To raise these questions is not to dismiss Tim and Bill's proposals. Rather, it is an effort to open up additional lines of critical attention and ethical consideration. While learning to recognize nonhuman subjectivities promises new modes of diplomatic engagement, it also introduces us to potential new forms of trauma and domination. Thinking responsibility in this context is complex, but it is necessary if we are truly to abandon outdated notions of "animal objects." Through the lens of recognition we are asked to consider forms of trauma that we cannot see, that we perhaps cannot even imagine; to ask not just about our effects on the material lives of others—their access to food and experiences of bodily pain—but also the subjectivities and socialities that our modes of engagement make possible. This is a process of empathetic attunement that is needy for its own forms of creativity. Possibilities for genuine forms of diplomatic response ask us to stretch ourselves in this way, to try, however impossible it might be, to imagine what something might mean to and for another. We inhabit a world of multiple and multiplying subjectivities. As such, the goal must be to multiply our own modes of recognition and response-ability for beings with diverse forms of liveliness, agency, intelligence, hopefulness, and more.

In doing so, we are also called into new modes of being. How will we be changed by the recognition of another as a being capable of looking back, by the knowledge that others are subjects who may be traumatized by our practices? What does it mean to inhabit a wakeful world, to be ourselves caught in the knowing gazes of so many others? Might this understanding present an opening into the humbling realization that there are countless modes of encounter and of recognition that we simply cannot, ourselves, recognize? How might these realizations transform us if we allow ourselves to be seized by, to become subject to, them? Ultimately, the question of recognition is that of who we will become, who we will enable one another to become, together. In posing these questions we are drawn into a practice that is about more than people doing conservation, one that must also ask seriously how we will be *redone* by that conservation (Maggs and Robinson 2016, 185). On reading an earlier draft of this chapter, Isabelle Stengers made a related observation: while some proposed conservation solutions provide creative new opportunities for collaborative experimentation, perhaps even for joy, she noted that others, such as mass killing, "endanger us."

Reflecting on the work of Hardshell Labs through this lens, I remain hopeful. At the very least, it must be noted that the oft-touted alternative to these proposals—large-scale shooting and poisoning of thousands of birds each year, in perpetuity—would not only kill these animals but also generate their own patterns of trauma, perhaps including forms of social breakdown. In other work, including other chapters in this book, I have explicitly refused to make choices between options simply because they're what's "on the table," when so much else is allowed to escape critical scrutiny as a result (see chapter 3 in particular). In this case too, any response to these ravens must be situated within a larger effort to address the diverse negative consequences of many contemporary modes of human life in the desert. But while that is happening—and there are at least some modest steps in this direction—these Hardshell Labs proposals seem to me to offer one of very few possibilities for a livable, even if thoroughly imperfect, way of inhabiting our current situation. Indeed, proposals like the techno-tort have the potential to reorient predator control away from killing, even away from ongoing harassment and policing, and instead open up the possibility that ravens might learn to live differently, *among* tortoises, in ways that everyone can accept. For this reason, on first encountering the suite of Hardshell technologies I was tempted to make a firm distinction between the learning potential of the techno-tort, on the one hand, and lasers and drones as technologies of harassment, on the other. But the reality is that we just don't know what these technologies promise or how they might be adapted. "We are," as Tim put it, "at the very beginning of a journey." In the years to come we may well discover that lasers and drones are equally technologies for something like learning: perhaps ravens will simply avoid them and population levels will adjust accordingly. Ultimately, however, I remain hopeful that at least some of the technologies being developed by Hardshell Labs will be or become creative, diplomatic proposals for genuine forms of flourishing cohabitation.

GIFTING

Crows gift. This fact, already well known to many close observers and feeders of crows the world over, found its way into wider public awareness in 2015 through the activities of an eight-year-old girl and her many crow friends in the suburbs of Seattle. Initially reported by the BBC, the story quickly spread, picked up by other news services and shared on social media (Sewall 2015). This young girl, it seems, had started feeding a large group of crows (*Corvus brachyrhynchos*) in her backyard each day, offering them tasty treats like peanuts and dog food. Of course, there is nothing particularly out of the ordinary about this kind of urban bird feeding. It was what the crows did that captured media interest: they started to bring a range of trinkets—from paperclips and buttons, to Lego pieces, an earring, and even a light bulb—leaving these objects behind on the feeding platform. Were these things gifts for this young girl, a kind of "thank you"? Were they perhaps something more akin to a payment? Were they not meant for anyone at all: random items carried and discarded, more or less purposefully, by one or two unusual crows?

Gift giving is frequently, perhaps always, shrouded in these kinds of ambivalences. Anthropologists and philosophers have spent—or donated?

FIGURE V.5. Illustration by Kirsty Yeomans (www.crowartist.co.uk).

perhaps gifted?—a great deal of ink exploring the precise nature of gifting relations. Questions of reciprocity, of (inter)dependence, of obligation, have often been at the forefront of their thoughts (Derrida 1995; Mauss 1970, 2002). When is a gift actually a gift? One thing that does seem to be common to many understandings of gifting is that it is a relational practice: it is about making and maintaining connections with others. Similar themes emerge in biological discussions. Where animals like crows gift things to others, usually their conspecifics, the behavior is generally understood to play an adaptive role in cementing significant relationships. Here too, questions often arise about the borders between altruism and self-interest, as well as the capacity for each of these possibilities to "masquerade" as the other (van Dooren and Despret 2018). Is a gift grounded in self-interest really a gift? What do we mean by "grounded" anyway: is evolutionary function or psychological motivation the more important factor?

Whatever the answers to these complex questions, the observable phenomenon remains: crows do seem to give things to others, which is to say that they "make another the recipient of something" formerly in their own possession (*OED* 1989). As Nicola Clayton (2015a) has noted, among corvids this behavior is particularly prevalent between mates during the breeding season, where it is "essential for maintaining the pair bond." But it is not unique to these relationships: "Crows are a social species and young birds will share objects as well as food to establish relationships with other birds, not just their partners" (Clayton 2015a). These sharing acts are made possible by sophisticated cognitive capacities. In their research, for example, Clayton and her colleagues have explored the abilities of Eurasian jays (*Garrulus glandarius*) to adjust which kinds of foods they give to their mates on the basis of what their mate is likely to prefer at a given time, irrespective of what the giver themselves would prefer (Ostojic et al. 2013). In the jargon of biology, these birds seem to be capable of "desire-state attribution," a facet of the possession of a theory of mind.

But in Seattle, as in many other similar cases around the world (Marzluff and Angell 2012, 108–15), gifting becomes the basis for new kinds of *interspecies* relationalities. Is it possible that, as with their mates, crows are paying attention to the *specific* desires of individual humans in these acts of gift giving? Of course, there are many instances in which this seems not to be the case at all, with crows offering—sometimes with remarkable persistence—unwanted and inappropriate gifts, including dead animals, in whole or in part. Another equally fascinating and disturbing example can be found in the writing of the biologist Konrad Lorenz, who tells of a tame jackdaw (*Corvus monedula*) who tried unsuccessfully, again and again, to feed him worms, eventually giving up on getting them into his tightly shut mouth and settling for stuffing them into his ear (1952, 129–30).[1]

In other cases, however, is there perhaps an attunement of some sort between giver and receiver, crow and human? Certainly, on the human side, many crow feeders are paying close attention to the needs and desires of their corvid friends and adjusting their gifts accordingly.[2] Are crows doing likewise, tailoring gifts to the imagined desires of human others? Studies of crows and crow feeders have identified complex "interspecific semiotics" in which *both* parties seem to be paying attention to, understanding, and responding to the other in sophisticated ways (Marzluff and Miller 2014). In Seattle, the mother of the little girl who fed crows reported one crow, captured on video, depositing in their birdbath a camera lens cap that she had lost earlier. In her view, this was definitely a purposeful act of gift giving: "They watch us all the time. I'm sure they knew I dropped it. I'm sure they decided they wanted to return it" (Sewall 2015).

We will likely never know the exact contours of the understandings and motivations that animate corvid acts of interspecies gift giving. This uncertainty, however, should not provoke an "escape" toward the seemingly sturdy, scientific-sounding explanations that all too often come into play in cases like this, explanations that reduce everything to adaptation and then eventually to evolutionary "self-interest." Even if corvid gifting has proven to be adaptive in cementing advantageous social bonds, this tells us nothing about what motivates individual crows in this activity or how it is experienced in its various expressions (van Dooren and Despret 2018). As Val Plumwood (1993, 143) argued in relation to human actions, black-and-white distinctions between self-interest and altruism are simply unhelpful. Most of our activities are wonderfully, productively, gray—in a broad range of different ways. Why should crow behaviors be any less ambivalent?

The social-media buzz that surrounded these gift-giving crows in Seattle is perhaps predictable. Part of me wants to ask why this possibility was so interesting, so surprising, to so many people. Was it the sheer fact that an animal might give a gift? Or was it perhaps the possibility for interspecies gifting? Either way, why are so many of us still so surprised by the creative, social, intelligent capacities of nonhuman others? But another part of me wants to hold onto the miraculousness of this act, to the image of a crow or perhaps a group of crows who might be paying attention to people who have done them a kindness, purposefully connecting through giving something that might be appreciated by another. Ultimately, I think that it was perhaps this possibility that most captivated people about this Seattle story: the possibility for engaged, mindful, reciprocal "connection" with a more-than-human world. Even if this is a mundane backyard occurrence, and one that in the age of the internet we are perhaps becoming less and less surprised by, I hope it remains an occurrence that we can continue to appreciate and be enlivened by. Crows remind us that our world is one that is—at least in part, at least some of the time—made up of and by diverse acts, ongoing traditions, of gifting between creative and thoughtful beings appreciatively reaching out toward one another.

FIGURE 5.1. Two aga (*Corvus kubaryi*) on Rota, Commonwealth of the Northern Mariana Islands.
Source: Photo courtesy of Phil Hannon and the USFWS.

Chapter Five

PROVISIONING CROWS

Cultivating Ecologies of Hope

ROTA, MARIANA ISLANDS

The first Mariana crow I encountered in the flesh was a young bird named Diseja. Perched up close to the mesh siding of her small aviary, she called out again and again as she saw us approach. It was not an alarm call, one of fear and warning at a potential threat; rather, she was begging. Having been orphaned and injured by a fall from the nest, Diseja had spent several weeks in captivity being cared for by a group of dedicated biologists. She had become habituated to their presence, even as they tried to avoid this, and lately she had seemed to ignore the other two Mariana crows (*Corvus kubaryi*)—called "aga" in the Indigenous Chamorro language—that were housed alongside her. As we got closer I was introduced to Diseja by Sarah and Phil, both biologists playing key roles in the conservation of this critically endangered species on the island of Rota in the Commonwealth of the Northern Mariana Islands (CNMI). Sarah Faegre had previously led the small team monitoring the roughly 150 remaining aga and was now contributing to this work while conducting research toward her PhD on the spatial ecology of the species. Phil Hannon, her husband, was heading up the recently established captive rear-and-release program that aimed to supplement, and hopefully expand, the small remaining population. It was June 2016, and we were in the backyard of a property rented by the San Diego Zoo to house staff working on the project. Eventually, the property would also be home to many captive aga: alongside the small working

FIGURE 5.2. Diseja, a captive aga on the Island of Rota. *Source*: Photo by author.

aviary that housed Diseja was the concrete foundation for a much larger aviary under construction by Phil and others.

Phil and Sarah explained to me that if everything went to plan, Diseja would be released long before this new aviary was up and running. Soon she would be taken to a release aviary in the jungle where she would not see people so readily and where other aga—some formerly captive, some not—were known to visit. She would adjust to crow community and be made ready for a life beyond the cage. At least they were hopeful that this would be the case, hopeful that the aga named "hope"—which is what *diseja* means in Chamorro—might soon be back in the jungle. Standing in the presence of this small, begging being, it seemed to me that her weighty name, provided by the two young girls living next door, might crush her. Hope. An act of naming that—perhaps unavoidably, however unintentionally—seemed to collapse the future into a single, fragile, living body.

Even given this introduction to aga, it was not until several days later that the significance of hope—of hopefulness and its absences—in the story of this remarkable species really began to become clear to me. Sitting in a diner called As Paris, one of the handful of restaurants on this small island, I was speaking with Thomas Mendiola, a deeply knowledgeable Chamorro man who lives and farms on Rota and has his own weekly radio show on Chamorro language and culture. Thomas and I talked about Chamorro cultural practices and about the history of colonization in this small Micronesian island chain: first the Spanish, then the Germans, then the Japanese, before finally the Americans. We talked about tourism in its many forms, about land ownership and the conservation of the island's biodiversity. But underlying and orienting all of this discussion, we talked about aga and the many ways in which the past and future of this species is caught up in the larger processes shaping life on the island.

Thomas was clearly an astute and attentive observer of crows. He told me about the way in which they could be relied on in the jungle, when hunting, to announce the presence of nearby animals of interest: "the crow is going to have a different cry for different animals," he said. Equally, on the farm, aga could be an ally, informing the farmer if a deer was in the field eating all the new shoots or if a cat was getting in among the chickens. "For the hunter and on the farm, the crow talks to you. It tells you about what's happening." He went on to say:

> I always enjoy watching the crow. This is a very smart animal as far as I'm concerned. This bird is interesting. I usually get my kids to come and see, because the crow is gonna open up these nuts and eat the insides. And the insides, you know, he doesn't eat it there. No, he starts to put it in the hole of a tree. Placing it in provision. This is a bird that makes provision! This is not a bird, you know? A bird doesn't make provision. This one does. It puts the nut in there then covers it.

Through this offhand remark, Thomas introduced me to a little-known behavior of the aga, its propensity to cache the nuts of the Pacific almond (*Terminalia catappa*), called *talisai* in Chamorro.

As our discussion continued, and in subsequent conversations with other people on the island, I learned that this behavior wasn't really "little known"—it was just selectively known. Although when looking through the biological literature on the species I was unable to find any mention of caching behavior, Chamorro people seemed to be well aware of it.[1] Biologists I spoke to had also frequently heard about this behavior from local people but had not themselves seen it. It seems, however, that there may be a very good reason for this: not only are there few crows left to observe, but the Pacific almond is itself a rare tree. But it hasn't always been. Thomas went on to tell me that the declines in both of these species are intimately entangled. Knowing that aga like almonds, knowing that the birds perhaps rely on these nuts or at the very least are more likely to hang around where they are found, many Chamorro people have taken to removing the trees wherever they find them.

Why? For the simple reason that they do not want aga on their land. Some people, like Thomas, clearly had a deep appreciation for the crow. Most others told me they didn't really care about the aga one way or the other: the birds had always been there, not particularly useful but not a problem, either. People and aga just kept to themselves. That is, until the efforts of the U.S. Fish and Wildlife Service (USFWS) and the CNMI government to conserve the aga began to interfere with local people's activities, in particular limiting land clearing and development in areas where the birds might be and, curiously, in some areas that everyone seemed to agree the birds had never been and weren't likely ever to be (Sussman, Ha, and Henry 2015). According to Thomas and many other people I spoke to, this situation pitted the aga against the people. In his words:

So there's this dilemma. It's created by the presence of the crow. Or by my wanting to be there [*laughs*]. Whichever one. The point is, which is it going to be: me or the crow? Now, the law that puts one against the other like that is a total confusion. If there's any danger to the crow, or *injustice* I would say, it is this law. If I am a crow, I'm going to tell the federal government, "Leave me alone. Why you bring this destruction to me? These people are my friends. Now they're looking at me like an enemy? Why make me an enemy. Now no one wants me in their place. People chasing me out."

Aga have been shot. They have been harassed and driven off land. They have had their food and nesting trees cut down, ringbarked, or poisoned. Put simply, they are now profoundly disliked by many. Not for any attribute or characteristic of the birds themselves; rather, in this case, they have become the epicenter of a complex and ongoing struggle over land but equally as importantly a struggle between different visions of the future of this island. In some cases their presence has very literally brought people's livelihood projects to a halt. But beyond the immediate and the tangible, these birds have become powerfully freighted with meaning, pervasively linked in the public imagination with all of what is preventing the development of this island, of its economy, and of people's standards of living.

Stan Taisacan, another Chamorro man, told me about people killing crows and about the ideas and attitudes that motivate this action. "The young ones [children] will hear others being so pissed off, embarrassed, that the crow is condemning the land. But actually it's never the crow, it's the human. It's the politics of it. It's so crazy." Stan too has a keen interest in aga. Retired now, he spent much of his working life as an employee of the local government, conducting biodiversity surveys. He was still upset by the personal price this work exacted, by the way in which he had been viewed as an enemy by other local people when conservation projects impacted, in his view often unnecessarily, on their lives. He argued against the listing of the species as endangered, telling biologists from the continental United States that the kinds of conservation measures they had in mind would just create tensions. He told me: "The crow coexisted happily with people until it became an issue and then boom, [the population went] all the way down. Is that how you say you're a conservation biologist? You've got to know the people first."

The aga's recent history is a story of relationships breaking down. Relationships between the people and the crow; between local Chamorro people, the government of the commonwealth on the island of Saipan, and U.S. federal agencies; as well as between crows and almond trees. Knowledge about crows that might have aided in better understanding them and their behaviors did not always flow easily between local people and conservationists, nor did ideas about how best to conserve them and to work with the local community. People who had previously ignored crows, or perhaps attentively observed them but largely left them alone, began to persecute and kill them.

Hope is my way into this situation. Underlying all of this activity are diverse forms of hopefulness, diverse visions for the future—many of them centering on the amorphous promise of economic development, principally in the form of agriculture and a rejuvenated tourism sector—which may or may not include a place for aga and other local species. Hope takes form here as diverse human imaginings and projects; "it comes to inform and shape the social practices of the present and form the contexts of immediately emergent futures" (Valentine 2012, 1051). But it also has distinctly more-than-human forms. In their caching of almonds, aga demonstrate a capacity for a particular kind of hopefulness, a particular kind of relationship with the future. Of central interest to me in these hopeful projects of both people and crows is the effort actively to bring about, to cultivate, particular futures, to take up hope as a work of "provisioning," to borrow and slightly repurpose this term that Thomas Mendiola drew into my thinking. In this context, as we will see, hope is a profoundly ecological possibility, one arising from, and only able to be sustained within, webs of enabling relationship. It is to these webs that we now turn.

LOSS

I arrived on Rota two days before my encounter with Diseja. Like everyone else, I had come via a neighboring island in the chain, either Guam or Saipan—in my case, Guam. There aren't really any passenger flights to Rota from anywhere else, a situation that in the days to come I would learn is viewed by many as a key part of the island's problems. Small and isolated, the tourism industry on Rota has all but collapsed. As I exited the airport I quickly found the rental-car desk. With only a handful of flights in and out of the island each week, it seemed that they had opened up the kiosk just for me. I filled out the paperwork and chatted with the woman at the counter while someone else brought the car around. As a first representative of the island, she was living up to the oft-repeated claim that Rota is "the friendly island." As the small red car pulled up to the curb it shuddered and almost stalled. I approached the man who got out of the driver's seat, who smiled and cheerfully said to me: "Don't worry about the sounds or any of the flashing lights on the dashboard. The car just needs a service. We were going to give it one but then you booked it." I soon learned that this is an island where almost none of the streets have names, so directions are given in relation to people's homes and other local landmarks (which were mostly lost on me); it is a place where drivers wave as they pass each other, with a sign on the highway encouraging people to do so; it is a place where fresh produce (for sale at least) is so hard to come by that making a salad requires stopping in at all five or six of the grocery/convenience stores on the island; it is a place where visiting researchers are encouraged to drop by the mayor's office and introduce themselves (which I did). In short, it is a small, friendly island, with all of the advantages and disadvantages, the comforts and the discomforts, that this affords.

Rota is the southernmost island in the Commonwealth of the Northern Mariana Islands (CNMI), with a population of roughly 2,500. The island of Guam is further south, and is part of the same chain but not part of the commonwealth. As a result of a convoluted history of colonialism, a political divide has opened up between Guam and the rest of the chain. Today, Guam is an "unincorporated territory" of the United States, while the other fourteen islands in the chain form the CNMI, a U.S. commonwealth (most of the population resides on the three southernmost islands). Different political systems, economies, sovereignties, patterns of development, migration, and militarization have accompanied these statuses. The CNMI's status as a U.S. commonwealth places it in a strange political position that implies some sort of sovereignty but that seems actually to amount to the kind of autonomy afforded a U.S. state, although sometimes without all of the rights. CNMI citizens have U.S. citizenship and the right to live and work in the United States. They can vote in U.S. presidential primaries but not in the actual presidential elections. The commonwealth sends a delegate to the U.S. Congress, but this delegate is unable to vote. The CNMI government is free to adopt its own laws in line with the 1975 covenant with the United States but is ultimately subject to U.S. federal law. This means that with regard to a broad range of issues, from immigration to trade agreements, key decisions are made in Washington. Guam has even less autonomy in local law making. It is an island dominated by two things: tourism and the U.S. military. There is, as Julian Aguon (2011, 65) has noted, a profound "psychic violence" in the relationship that the United States has with Guam—and in a distinct but related way with the larger CNMI—in which local people "must find our way in a country that neither wants us nor wants to let us go."

I am wary of narrating this as a history in which local people have simply been pushed around by larger, external forces, a storyline that local scholars have noted dominates histories of this region and does a great deal to suppress important forms of agency and resistance (Diaz 1994, Delisle 2015). But the many layers of occupation and colonization in the islands are centrally important to their shaping and have left profound and lasting legacies (Camacho 2011). The Chamorro people settled in the islands roughly 4,500 years ago, coming from Southeast Asia. Like other Micronesian peoples, they farmed and fished, and in the Marianas they crafted distinctive *latte* stones (which may have been used as supports for buildings). The island chain entered European history in 1521 with the arrival of the Portuguese navigator Ferdinand Magellan. Roughly one hundred years later the Spanish colonized the islands, violently suppressing and relocating the Chamorro people in an effort to quash ongoing resistance. In the 1890s the Spanish lost Guam to the United States in the Spanish–American War and subsequently sold the rest of the island chain to Germany, opening up a division between these islands that would grow in the coming decades. During World War I, Japan took possession of the Northern Mariana Islands, its claim later being officially recognized under the terms of the Treaty of Versailles and subsequently by the League of Nations. With the onset of World War II, Japan seized Guam too (from the United States), violently suppressing Guamanian Chamorro resistance (Camacho 2011). But this state of affairs was short-lived, with the United States invading all of the Japanese-occupied islands in the Marianas in 1944. The United States subsequently established several military bases in the islands from which to conduct the bombing of the Japanese home islands; the atomic bombs dropped on Hiroshima and Nagasaki took off from the Mariana Islands.

Today, the U.S. flag continues to fly over these islands. For the Chamorro people, each wave of colonization has brought different changes and challenges, many of them with long-lasting legacies for both the land and the people. As Robert Ulloa, a Chamorro man and employee of the CNMI Division of Fish and Wildlife, summed it up for me:

> With the Spanish, they were not much concerned with conserving our resources, they were more interested to establish religion. Then came the Japanese, they were not focused on religion or cultural preservation; they were here for industrial reasons. Now we have the American colonization which is more focused on preservation of both heritage and natural resources.

Of course, there are many other important facets of the U.S. influence on these islands, but conservation is certainly one of the key ways in which local lives are shaped—for good and bad—by this most recent colonial power. To date, in Rota at least, the aga has been one of the most significant examples of these dynamics.

Originally found on both the islands of Guam and Rota, the Mariana crow is today completely extinct on the former and just clinging to existence on the latter. The aga is a smallish omnivorous crow, its diet composed of a wide range of plants and animals, including insects, smaller reptiles, and bird eggs (USFWS 2005, 17). They are forest birds; relatively shy in nature, they tend to keep within or below the canopy rather than soaring above. It is likely that the Rota population was always smaller and less genetically diverse than that on the much larger island of Guam. The cause of the species' disappearance on Guam is relatively straightforward. As with so many other animal species, predation by the brown treesnake (*Boiga irregularis*) is thought to have been the primary factor. First reported on the island in 1953, the snake was accidentally introduced to Guam in the 1940s, a stowaway among cargo being shipped through this central node in the U.S. military's immense Pacific network. The species quickly spread from south to north across the island, consuming and ultimately wiping out a wide range of species as it went (National Research Council 1997). "Birds, bats, and reptiles were affected, and by 1990 most forested areas on Guam retained only three native vertebrates, all of which were small lizards" (Fritts and Rodda 1998, 113). The Chamorro poet Craig Santos Perez (2014) has captured the entwined processes of colonization and extinction at work here in the simultaneous displacement of people and birds:

> the snakes entered
> without words when [we] saw them it was too late—
> they were at [our] doors sliding along
> the passages of [*i sihek*]
> empire[2]

By the 1980s, the aga was completely gone from most of Guam, with only a single, tiny population hanging on in the north. The primary impact of the snake seemed to be on eggs and nestlings, so efforts were made to develop electrical barriers on and around remaining aga nesting trees to keep the snakes away. A few crows were even brought from Rota to supplement the flagging population. Ultimately, this activity was to no avail. The last sighting of an aga on Guam occurred in 2011.[3]

The decline of the species on Rota is a bit more of a puzzle. The brown treesnake has not yet established itself on the island, and there is a significant monitoring and trapping effort in place to prevent it doing so.[4] This situation is also helped by the fact that Rota is "off the beaten track," as it was put to me by Renee Ha, the University of Washington biologist who has overseen much of the aga monitoring for the past decade or so. With many fewer flights and freighters than other nearby islands, there is simply a lot less opportunity for stowaways. More generally, Rota's being "out of the way" has meant that it has been spared much of the ecological destruction that conflict and major building projects have wreaked on neighboring islands. Importantly, this includes the fact that Rota was not heavily bombed or extensively deforested for military use during World War II, as other islands in the chain were and indeed still are. As Thomas Mendiola put it to me: "Rota is more untouched. For whatever reason. They don't have interest in it." Habitat loss as a result of agriculture and land development for industry and tourism has also been substantially less significant on Rota than on the other inhabited islands in the chain. But even in this far less disturbed environment, the aga has declined precipitously.

When exactly the aga population on Rota began this decline is a matter of some contention. Surveys were infrequent, and some of their findings have been questioned. Studies seem to show a massive decline in the late 1980s and early 1990s, from a population of roughly 1,300 birds to fewer than 600 (National Research Council 1997, 35) and down significantly further since. Some uncertainty also remains around the causes of this decline, with biologists generally pointing to a few key factors. Predation, especially by domestic cats, and in particular of juvenile crows, is thought to be central. While older birds are more savvy about predators, a high number of aga seem to be dying in their first year, a time in which they are awkward, inquisitive, learning to fly, and spending more time than they would in later years on the jungle floor (Zarones et al. 2015). In an effort to address this situation an extensive cat-trapping program has been established throughout the aga habitat areas. During my trip I had the opportunity to meet with Doug Page, who runs this program, from the Institute for Wildlife Studies, and I joined him on his daily routine, which involves several hours of driving and walking to check each of these traps. Alongside predation, conservationists also point to the possible impact of habitat loss on the species, especially during earlier periods of more extensive development on the island, and that caused by—perhaps increasingly common and severe—typhoons. At the time of my visit there was also speculation that an unknown disease might have long been affecting the species; more recent research indicates that some birds are dying of liver and lung failure, but the precise cause is still unclear. Finally, the drastic narrowing of the species' gene pool may have led to inbreeding depression and lower nesting success. All in all, as Sarah put it: "There are a lot of unknowns. Those are all issues, and there may be other issues that we don't understand yet. It's really sort of a question mark as to why they've declined so steeply in recent years."

Alongside these interwoven factors, as we have seen, it seems that deliberate disturbance and persecution of crows by local people has played a role in their decline. This kind of activity has always taken place to some extent on Rota, with local people sometimes shooting or driving off aga who were known to steal chicken eggs and perhaps cause other kinds of nuisances (National Research Council 1997). But beginning in the 1990s or early 2000s, people told me that the targeting of aga was drastically scaled up. It is not possible to quantify the extent of this activity with any certainty, but without exception locals said that they knew about people

FIGURE 5.3. Farmland on the island of Rota, Commonwealth of the Northern Mariana Islands. *Source*: Photo by author.

killing aga and/or removing their habitat in a targeted manner, and most of them had seen it or taken part in it themselves. Several, like Thomas and Stan, claimed that this was the *principal* cause of aga decline on Rota.

At the heart of local animosity toward the aga is the way its conservation has held up land clearing for farming or development. On Rota, as in the CNMI more generally, only Chamorro people can own land. Article 12 of the CNMI Constitution "limits permanent and long-term (more than 55 years) land acquisition to people who are at least one-quarter Northern Marianas descent and corporations that are 100 percent Northern Marianas owned" (Ristroph 2007, 39).[5] Chamorro people are allocated a homestead for farming through a lottery. The particular area of homestead lands at the center of the aga controversy is located in the eastern part of the island, around an area called Duge. As a result of miscommunication or perhaps deliberate political strategy on the part of a past local mayor facing reelection (I heard various stories), these lands were parceled up and people told about which they were to receive before an environmental impact statement had been completed or formal approval issued by the CNMI government. People began clearing "their" land until they were ordered to stop. This situation reminds us that contestations over the aga are not simply a matter of U.S. federal involvement in local lives. Frequently, as in this case, disagreements also involve complex questions of funding, permitting, and politics that engage the commonwealth government on Saipan and the municipal government of the island of Rota.

Up until the time of this controversy, Stan explained to me, the local people "don't know the crow, don't know anything about it." Here began a long and very drawn-out process. While efforts were made by all levels of government to develop a Habitat Conservation Plan—a legal mechanism within the ESA that might have allowed other crow habitats to be set aside in compensation for the jungle lost to homesteads—this never quite worked out. Year after year of delay bred resentment in the community, creating what one biologist referred to as a "festering sore." Ultimately, homestead lands allocated in the 1990s were only freed up for use in 2014, when an agreement was finally reached between the various government agencies. For the intervening decades, many people were unable to build family homes or begin farming. As one local man put it to me: "This is a long time not to have a livelihood." Several people told me that more than one person had died waiting for their land.

It is hard to overestimate the significance of this situation. As Robert A. Underwood (2001), a former Guam delegate to the U.S. Congress and president of the University of Guam, put it, land is the "singular issue that can radicalize even the most mild-mannered Chamorro from the loyal military retiree and the police officer to the teacher and the nurse." And so the change that occurred in 2014 may have been too long in coming. This change was the result of an agreement in which the Rota government set aside a large area of coastal forest as an official Mariana Crow Conservation Area. People with homesteads in the Duge area were allowed to begin clearing and building, provided that there were no crows nesting on their site. If there were, they had to wait until the end of the breeding season.[6] This is a lower bar than usual on the island, where any clearing of more than 100 square meters requires permits from the CNMI government and any activity that would affect the aga or another listed endangered species triggers the involvement of the federal USFWS. If this impact cannot be avoided, any development then requires the issuing of an Incidental Take Permit (ITP) and the drafting of a Habitat Conservation Plan (HCP), which can involve considerable expense and delay for the landowner (even with the dedicated work of many people I spoke to in these agencies who are working to reduce these impacts).[7] While these conditions have for now been relaxed for the Duge homesteads, they remain in place everywhere else on Rota and are likely to continue to cause frustration.

In my conversations with locals I also encountered frequent misunderstandings about this new arrangement, with several people telling me that the crows now "have their area" and so there shouldn't be any more need for conservation restrictions on other land. There is, it seems, another potential site of disquiet brewing here, especially given that many other species are now joining aga on the endangered species list (in particular from late 2015). As Lainie Zarones, then the CNMI's Endangered Species Program Manager, pointed out to me in an interview in 2016, Rota now has "a higher concentration of endangered species than anywhere else in the United States." This means that permits, delays, and additional expenses associated with development are only likely to increase in the coming years.

People that I spoke to on Rota provided very different pictures of local responses to this bureaucratic process: some said that most people applied for permits and took their chances, others claimed that almost no one did, relying instead on surreptitious acts of clearing or vegetation poisoning. Either way, one thing was clear: things went more smoothly if there weren't any crows around. Thomas explained to me (taking the position of a hypothetical person in this situation): "Me, who wants to build my house there, I have to find a way to get rid of the crow. Of course I'm not going to shoot it, that's too drastic. But some people will. Nobody is watching them. They'll do it. It's illegal. But for others, we know the reason why this crow is here. Why? Because there are these almond nuts right there. Cut that off. Watch the tree fall down and no more. Crow gonna go away. And now I can get my land."

As previously noted, according to many local people this has been one of the key causes, if not the major cause, of the aga's decline. As Robert Ulloa explained: "Aga and Chamorro long coexisted. People were never really in the mindset of harming the birds until the preservation regulations were introduced. People were just bombarded with regulations. And so they see it as unfair for their livelihood that aga should come before the people." Some Chamorro people are openly dubious about the effect of cats and other predators on aga. Similarly, few people that I met thought that clearing or farming might have any real impact on the species; they told me that their ancestors have lived this way for the longest time. Others said that perhaps these factors are relevant *now*, especially with the small remaining population of aga, but that direct persecution and the removal of specific trees are still the key causes of the decline of the species. In this way they claimed a kind of ownership of this (pending?) extinction, not so that they might take the full weight of responsibility onto themselves but so that it might be directed at conservation outsiders and government agencies, both commonwealth and federal, for the way in which they have handled the protection of this species.

If these claims are correct, then the aga wasn't declining on Rota when the government began to conserve it (although it certainly was on Guam, but because of the presence of the brown treesnake). Instead, the act of conserving, grounded on uncertain local survey data, may have set in motion a chain of events that has itself pushed the species to the edge of extinction. Stan explicitly told me that this is what had happened, calling the current situation a "cover up." On the other hand, however, the aga may already have been in decline to some extent on Rota too, from predation by cats, habitat loss, and perhaps even an unknown disease. Through my interviews and reading of the relevant literature, I haven't been able to untangle the many timelines and separate out the many causes of decline with enough certainty to know what happened here. Indeed, I don't believe that it is now possible really to do so. However it happened, today we have arrived at a situation in which the future of the aga is deeply uncertain. As one biologist succinctly summed things up to me: "That bird's going to go extinct in our lifetime." One thing that is clear in the midst of all this uncertainty is that the particular mode of conservation taken up was less than ideal. This should not come as a surprise to anyone; the USFWS have themselves acknowledged the many drawbacks of the "one-size-fits-all" legislative requirements that they work within in cases like this.[8] Here, in a range of different ways, local people felt that their needs and understandings were not adequately considered. As such, whatever the starting point—and of course, there is no single starting point for anything inside long tangles of association—the result was the same: the relationships that had enabled Chamorro and aga to coexist on this small island for so long broke down.

POSSIBILITIES

About a week into my trip I found myself driving into Songsong, one of the two main villages on Rota. Something of the shock that I had experienced on first entering the village still lingered, but I was struck by how familiar the place had become. Rota is a landscape of contrasts, of beauty amid and often in ruin. Wherever I looked I saw abandoned buildings. Almost everything on the island is built from concrete; the occasional strong typhoons that move through flatten any other kind of structure. All around me were old houses, many now without their roofs, concrete shells of homes. Perhaps a third of the houses in the main part of town were abandoned. Large commercial buildings were also empty, many of them in better states of repair but faded and broken down nonetheless. As I drove out the other end of town on my way to West Harbor, I passed the island's small port facility building and the remnants of one of the only buildings not made of concrete I'd seen: a massive shed, collapsed and now burying a huge, heavily rusted crane within it. But alongside this disrepair, it was also becoming increasingly clear to me that in other ways Rota is a spectacularly beautiful and well-cared-for place. The coastlines

FIGURE 5.4. An abandoned house in the village of Songsong, Rota. *Source*: Photo by author.

FIGURE 5.5. The spectacular coastline of Rota. *Source*: Photo by author.

were like nothing I have ever seen before: in some places sheer cliffs with abundant bird life, in others, jungle spilling out onto beautiful beaches surrounded by shallow coral reefs. Picnic areas dotted the highway around the island, many of them well maintained and frequently used, part of the new mayor's highway-beautification project.

Traveling around the island I was reminded that hope is a proposition that is much easier to grab hold of, to put to work, in some places than in others (Kirksey, Shapiro, and Brodine 2013, Tsing 2015). Standard definitions of hope tie it to wants and desires for the future, that something good or at least better might come about. Exploring the island of Rota, drawing on conversations and reading, I was slowly coming to understand the way in which people's hopes were tangled up with the economic decline so visible around me. As with the rest of the CNMI, in the late 1980s and early 1990s Rota experienced a period of significant economic growth. Driven by Asian investment, a strong tourism sector (especially catering to nearby Japan and South Korea), and a booming garment industry, the rate of economic growth in the CNMI "was one of the highest of any American state or possession at the time, and higher than the growth rate of the 'Asian Tigers'" (CEDSPC 2009). But all this began to change in the mid-1990s. There was no single cause. Tourism was hit hard by a series of factors: downturns in key Asian economies, legal disputes over some foreign-development proposals in the islands, and in the early 2000s a general reduction in travel as a result of increases in fuel prices and a series of international and regional events, including the September 11, 2001, U.S. terrorist attacks, the SARS epidemic, Bali bombings, and Asian tsunami (CEDSPC 2009, 12). Perhaps the single largest factor in the declining economic fortunes of the CNMI, however, was the almost complete disappearance of the garment industry. Following the liberalization of U.S. import rules for clothing made in China, the commonwealth lost its competitive advantage—an advantage that was already being undermined by U.S. federal intervention into the substandard sweatshop conditions that long existed in the CNMI (Shenon 1993, Brooke 2005a). In a period of just a few years, an industry that comprised a third of the commonwealth's economy and one of its largest sources of employment completely shut down (CEDSPC 2009). On Rota, declining interest from tourists was paired with further changes to airline conditions, which reduced the number of flights and made connections from Guam and Saipan more expensive and time consuming (a major factor for many time-poor Asian tourists). From several passenger flights into the island a day, they were now down to a few per week.

The whole commonwealth continues to struggle economically. According to a strategic plan developed by the CNMI Department of Commerce: "By definition, the CNMI is in an economic depression unmatched even by the Great Depression of the United States in 1929–1933" (CEDSPC 2009, 7). Thousands of people—immigrants, Chamorro, and other locals—have left the islands for the continental United States or elsewhere in an effort to find employment and other opportunities for life. From a population of roughly 68,000 in 2001, today the CNMI is home to about 52,000, a massive and globally quite unprecedented 23 percent decline (U.S. Census Bureau 2015). The vast majority of the population lives on the island of Saipan, with only a few thousand people on the other two main inhabited islands in the commonwealth, Rota and Tinian. Rota's current population of roughly 2,500 people represents a reduction of more than 20 percent over the past two decades. This, I was told, is one of the key reasons for all of the dilapidated properties: their owners simply aren't around to take care of them. And even if they are still on island, they're not always in a position to do so. Financial assistance from the United States now accounts for a huge proportion of the CNMI's revenue, over 35 percent in 2013 (CIA 2016). The tourism industry limps along, a shadow of its former self, but now firmly positioned as the largest employer and revenue producer in the commonwealth.

With this in mind it is not surprising that tourism also dominates many people's visions for the future of the island, although, among the people I spoke with there were many different ideas about what that tourism should look like. When I asked Robert Ulloa if he thought local people were hopeful about future growth in tourism, he replied that they were not. He went on to say: "A lot of people are living off island to try and find better opportunities. You can see that a lot of our buildings here are left abandoned and not maintained. At times people would come into our village and see it as a ghost town because people are not out on the streets. So that hope is not within the people." As our conversation continued it became clear that Robert felt that people would like tourism to return, would in fact love for this to happen, but that many of them had given up on it. In making this distinction, Robert reminded me that a hope is not simply a desire. In order to hope, one must also believe that one's desires have a chance of being realized, however slim. Otherwise we *lose* hope; it slips through our fingers.

There is a particular kind of relationship with the future at work here, a distinctively liminal one. Hope resides between certain failure and success, between despair and delusional optimism, between reality and the ideal. Each of these borderlands matters. If a particular future is impossible or, on the contrary, if it is certain to arrive, then either way it cannot really be said to be "hoped" for. Precisely because of this uncertainty, hope also sidesteps both despair and delusion; while it can certainly slide toward these extremes, its fickle nature always pulls it back from the edge. Similarly, hope is only possible in so far as reality "fails" to achieve an ideal: if we lived in a perfect world there would no longer be anything to hope for.[9] This is an understanding of hope that refuses a simple contrast with "despair," a hope that cannot be reduced to "optimism." Instead, hope is a deeply mortal proposition: it is a future-oriented mode of inhabiting a fundamentally uncertain and imperfect world, yet one in which something better is possible.

But there is another important, ambivalent specter looming over the Mariana Islands, a source of both tremendous optimism and great anxiety about the future. The specter is that of the U.S. military. Guam is today home to significant U.S. Navy and Air Force bases, facilities that together directly control roughly a third of the island. In recent years these bases have taken on increasing importance as geopolitical tensions heat up in the Pacific in relation to both contested territories in the South China Sea and the nuclear ambitions of North Korea. What's more, this significance will likely only increase in the coming years, as the major U.S. military presence in Japan winds down, part of a massive planned redeployment of troops from Okinawa to Guam (Aguon 2011, Weaver 2010). Local people are increasingly finding themselves caught up in the consequences of this military expansion in the Pacific, a situation that burst onto the world stage in August 2017 when North Korea threatened to take out its frustrations with the United States by launching a missile strike on Guam (Berlinger 2017). While Guam lies outside the CNMI, the growth of the U.S. military presence there will have a range of significant consequences for Rota and the rest of the commonwealth. It will likely suck more people out of the CNMI, attracted to new jobs in Guam both in the military and in various service sectors. But at the same time, it is expected by some that a growing population on Guam will create new tourism and other economic opportunities, with the islands of the commonwealth providing "a variety [of] activities not available in Guam, inclusive of gaming on the islands of Tinian and Rota, as well as eco-tourism on the island of Rota" (CEDSPC 2009, 8). (Gaming is no longer really available on Rota, with the casino having closed down. During my visit, however, rumors were circulating about an even bigger casino/resort proposal.) Importantly, traveling alongside these developments is a heated contestation over the current and future use of previously uninhabited CNMI islands, like Pagan Island, for destructive weapons testing and live-fire training by the U.S. military (Cave 2015; Hadfield and Haraway, in press).

The CNMI government's own strategic economic plan presents the buildup of a U.S. military presence in the area as something that will "undoubtedly bring both positive and negative challenges." But "faced with many economic challenges the Commonwealth looks to an opportunity which is already bringing in new investment to the region" (CEDSPC 2009, 7). In this regard, this small commonwealth is not unlike many of its people. For the people too, the military is often seen as a less-than-ideal option in a space of limited choices: it is one of few viable routes for local people to both continued education and off-island mobility. Rates of enlistment are very high here, as they are in many other parts of the American Pacific where incomes are low. "While small in real terms, enlistments from Guam, Saipan, and American Samoa are the nation's highest per capita." Taking advantage of this situation—of a kind of hopelessness—"the Army has found fertile ground in the poverty pockets of the Pacific" (Brooke 2005b). As one local put it to me on Rota: "The military loves places like this. They come along and just vacuum up the youth."[10]

In this place of limited opportunities, one thing that people have is land. Whether to build a house, to farm, or to put under a long-term lease to a developer, land is an important source of livelihood. Understanding this broader context is crucial to understanding the aga story because, of course, it is precisely these vital relationships with the land, these possibilities, that the crow is now seen to threaten. This situation was borne out clearly for me in a discussion with two Chamorro men, both holding senior roles in natural-resource management on Rota. I approached the men and their team while they were hard at work in a park by the beach on the edge of Songsong, rebuilding some storm-damaged huts that are used for community events. They agreed to take a break to chat with me by the water. One of them summed the situation up simply: "What is for the future of the islands is development. We can't expand development because of these issues [conservation, principally the aga]." The other added: "Our development has been held back because of these issues." To which the first replied, "We're living in a primitive age again." In this way, aga is deeply entangled—passionately and mortally caught up—in hopes for the future. While, as Paige West (2006) has argued, around the world conservation and development are now more and more being drawn into strange alignment—through a range of often profoundly unequal "conservation-as-development" projects—here on Rota they are still more commonly seen as opposing forces.

But this too is changing. In conversations with locals, ecotourism was frequently floated as a way of reconciling competing demands. People were well aware that the island had healthier jungles and reefs, that it had more biodiversity, than neighboring islands. In fact, the key area of tourism that seems to have survived is scuba diving. Some effort has been made to provide basic amenities and sites for these kinds of tourists: two walking paths have been cut through the jungle and a bird sanctuary established, including a spectacular lookout where visitors can stand at the top of a coastal cliff and watch a range of forest and seabirds nesting and playing in the updrafts. Like so many other things on the island, while it is obvious that great care has been taken over some aspects of this sanctuary and its maintenance, others have slipped away—the signage is now weathered to the point that it can't really be read. But this is certainly the direction that many people are now imagining for the future of the island. Driving past the airport on the way to the bird sanctuary, a sign stretched across the highway announces: "Welcome to Rota: Nature's Treasure Island." In my brief meeting with the mayor, he too emphasized the importance of ecotourism for Rota, the way in which it might contribute to the economy and take advantage of the island's unique attributes and heritage.

As in all such dreams of the future, the devil lies in the details: which projects will be taken up, and at what cost to whom? While generally supportive of ecotourism, Thomas Mendiola wondered whether it was really what people visiting the Marianas were after. His experience led him to believe that many tourists had a pretty narrow band of interests: "sun, sea, and sand" as he put it. Neither natural heritage nor cultural heritage seemed to matter to them in anything other than a superficial way. This view was borne out for me by the prominence of Hawaiian and other Polynesian dancers and cultural activities in the tourist centers of Guam, staged activities imported from elsewhere to satisfy tourists' desires for an iconic "tropical island experience." This is perhaps a contemporary example of what Elizabeth DeLoughrey (2017) has referred to as "*aqua nullius*," in which the tourist's narratives and desires empty out and redo other people's sea-/landscapes in ways that are both symbolic and thoroughly material. The actual richness of local cultures and places is here sacrificed for the sake of the imagined ideal (Peterson 2016). Stan Taisacan, on the other hand, doubted whether the "eco" part of the tourism would actually mean much for long. He pointed to the way in which the island's I'Chenchon Bird Sanctuary, originally imagined by him as a very basic affair, now included a parking lot too close to the birds, tables and benches, and a toilet block. He worried that the facility might now compromise the site, jokingly telling me about his concern that one day they would build an elevator down to the birds. Meanwhile, some biologists wondered whether new tourist ventures would be regulated in a way that required them to contribute to, rather than destroying more of, the environment. For example, they asked, would developers be encouraged or required to renovate some of the many abandoned hotels now dotting the highway, being slowly reclaimed by the jungle, or to take care of the now abandoned pool and water park? Or would they be able to take the cheaper and easier option of simply clearing more land and occupying more of the coast? In short, many questions remained to be answered about the forms that even this more hopeful and inclusive of "eco" futures might take. While aga and other endangered species might potentially take on a new value, they might just as easily be squeezed out.

We are reminded here that hopes take varied and often conflicting forms. The futures that some of us seek will not accommodate others—other living beings or their possibilities. Far from being an innocent or inherently life-affirming orientation, acts of hopefulness are unavoidably caught up in larger patterns of unequal world making. Whose hopes are realized, and at whose expense? Who gets to hope at all? If, as I have argued elsewhere, responsible hope is a kind of care for the future, then that care must simultaneously be a process of interrogating the many

FIGURE 5.6. An abandoned pool and small water park just outside the village of Songsong, Rota. *Source*: Photo by author.

other futures that are rendered impossible or unlivable through our particular practices of hoping. Drawing on Maria Puig de la Bellacasa's (2012) sense of care as simultaneously "a vital affective state, an ethical obligation and a practical labour," hope becomes a fleshy, thick possibility. Such an understanding of hope works against its romanticization, instead insisting that we take up the difficult work of interrogating what is hoped *for*, and with what consequences for whom (van Dooren 2017). On Rota, there are many possible futures bubbling away. It remains to be seen which of them will rise to the surface and what forms of life and livelihood they will render possible and impossible for local people, crows, and others.

PROVISIONING

On my last weekend on Rota I went aga tracking with Dacia Wiitala, a biologist from the United States who had worked on the project for several years and had recently taken over the lead role in the monitoring team. Beginning at 6 a.m., for about four hours we trekked up and down hillsides, through dense jungle and grassy clearings, following the mysterious beeps of a radio telemetry receiver in pursuit of tagged crows.[11] I enjoyed the opportunity to see more of the aga habitat, to get a feel for where these crows live, even if we didn't see much of the birds themselves. We found Groucho, the first bird on our list, after about forty minutes of searching, aided greatly by Dacia's familiarity with his patterns of movement and usual haunts. Now healthy and doing well, Groucho had been given his name because of the odd pattern of baldness on his head when he had first been encountered and banded as a juvenile, giving him the appearance of two large, bushy eyebrows (which reminded the researchers of Groucho Marx). Through dense branches we could just make him out, sitting in the mid canopy. With binoculars Dacia confirmed that this was indeed Groucho (with the aid of his color-coded leg bands), and I managed to take a couple of quick photographs before he took off again, cawing as he went.

The second bird, Roy, proved to be much more problematic to locate. She led us around in a large circle, through dense jungle and almost impenetrable thickets of hibiscus. At a couple of points we walked through incredible karst forest, where the roots of tall trees spread out into uneven

FIGURE 5.7. Groucho, spotted while aga tracking on Rota. *Source:* Photo by author.

floors of limestone rock with no visible soil at all. This limestone, the afterlife of ancient coral reefs, had been forced up by geologic activity and now covers most of the island (NPS 2005, 6). In patches like some of the ones we walked through, this history is readily apparent, with shells and imprinted coral shapes scattered throughout. As we walked, the shifting rocks beneath our feet made a sound like terracotta pots scraping against one another. For about two hours we used the radio receiver to follow Roy in this way, occasionally hearing her but never able to get visual confirmation. It is visual confirmation that monitoring trips like this aim for, ensuring every two days that the birds are alive, in part so that any dead ones can be collected for autopsies before the jungle consumes them. Eventually, Dacia decided to halt the search, concluding that Roy was likely socializing instead of foraging (thus the erratic movements) and that the occasional calls we had heard would have to suffice for the day.

As we walked and drove that morning, covered in sweat from the already hot and humid day, we talked. Having heard for the first time earlier that week about agas caching Pacific almonds, I asked Dacia if she had seen this behavior in her hundreds of hours spent observing the birds. She hadn't. I asked if she had ever heard about it from locals. She had, from a range of different people. In fact, she told me, people had told her that they had once raided these stashes of tasty nuts to eat them themselves, something that Thomas also mentioned. Dacia hoped that she might one day see birds caching almonds, too, but, she added, there were now few of the trees left on the island, at least as far as she was aware.

Through this fascinating caching behavior we learn that on Rota the crows are working toward particular futures, too. Ecotourism and casinos are likely not a part of what they have in mind, but they are nonetheless imagining what might yet come to pass. Cognition is an elusive thing—even in other humans—but experimental work does now enable us to say with relative certainty that crows are among those animals that imagine and work to bring about specific possibilities for the future. Fascinatingly, in research all over the world, the way in which various species of crows cache food items—"provisioning," as Thomas Mendiola put it—has been at the center of scientific efforts to better understand this "prospective cognition."

The notion that crows might have ideas about the future works against a tide of older biological thought. The Bischof-Köhler hypothesis proposes that "animals other than humans cannot anticipate future need or drive states, and are therefore bound to a present that is defined by their current motivational state" (Suddendorf and Corballis 1997). In line with a long and rich tradition of "anthropocentric propers"—capacities imagined to be uniquely, and perhaps definitively, human (Routley and Routley 1979)—from this perspective, (other) animals are trapped in their present reality, unable to imagine or take action to influence a future need or desire. Of course, we see animals doing all sorts of things that *look* future-oriented: why else would a bird go to the bother of constructing a nest, if not with the idea of eventually laying eggs in it? But for the behavioral biologist, these kinds of activities might just as easily be the result of a "fixed action pattern" that is triggered by hormones or by changing temperatures or light (Raby and Clayton 2009, 315). In short, the bird may have no clear sense of *why* she is doing what she is doing: she may know how to build a nest, and she may even be able to modify that nest to better suit local conditions, but does this mean that she knows it will one day become a home to eggs and chicks? With this in mind, biologists tend to distinguish between behaviors that are grounded in a "sense of the future" and those that, while seemingly future-oriented, are not (Raby and Clayton 2009, 315). Within that category of activity *with* a sense of the future, a further distinction is frequently made between (at least) two possibilities, referred to as "semantic" and "episodic" thinking. If our nest-building bird does have a sense of the future, does she know, in a relatively abstract way, that eggs might one day be laid here (semantic), and/or is she able to imagine *herself* in that future, perhaps laying and incubating eggs and rearing chicks (episodic)? Debate continues about how to make these kinds of distinctions and about which of these various modes of thought might have evolved first or might be more "advanced."

Corvids, with their propensity for storing and then recovering various items—usually food—have wandered into this debate and taken center stage. Caching is a fascinating behavior. While relatively widespread among mammals and birds, the corvids seem to have a particular talent for it.[12] The behavior starts very early in life, and there seem to be compulsive dimensions to it (Grodzinski and Clayton 2010, 977). Others have suggested that there may be an element of joy, too: crows, jays, and other corvids may really like hiding things away to collect at a later date (Bugnyar, Stöwe, and Heinrich 2007, 758). As Nicola Clayton, one of the world's leading biologists in this area, explained to me in an interview, caching has been such a fascinating behavior for biologists because it lends itself so well to experimental inquiry. While it may well be that in building their nests, birds are relying on sophisticated memories of the past and imagined future scenarios, it is very hard for this to be tested experimentally.[13] But when corvids hide food items, and again when they return to collect them, researchers can explore the extent to which they understand their actions in any given present as either making possible future activities or as enabled by those of the past.

Many of the key studies on prospective cognition have been conducted in Nicola Clayton's lab at the University of Cambridge. These studies have focused on the western scrub jay (*Aphelocoma californica*) and the Eurasian jay (*Garrulus glandarius*), close relatives of the crows and members of the larger family Corvidae (Raby et al. 2007; Correia, Dickinson, and Clayton 2007). Related studies in Thomas Bugnyar's lab at the University of Vienna have focused on common ravens (*Corvus corax*) (Bugnyar et al. 2007, Bugnyar and Kotrschal 2002, Bugnyar 2010). In the absence of similar studies on the aga—the conditions for which no longer really exist—we can have no definitive evidence but might quite reasonably postulate that similar capacities and understandings lie behind their similar behaviors. As Nicola pointed out to me, the results of research on all of those corvid species from around the world that have been studied—in this and other areas of cognition—have been "remarkably convergent."

In what was perhaps the first key study on corvid prospective cognition, jays were familiarized with two different rooms that they might be allowed access to each morning. In one room they were given breakfast but in the other room they would go hungry. When provided with access to food and both rooms in the afternoon, birds preferentially cached in the room where they could reasonably expect to be left hungry in the future. To avoid the possibility that caching behavior was unthinkingly "triggered" in this place because of a *past* association with hunger rather than the anticipation of possible *future* hunger (which in itself might be an interesting indication of memory/retrospective cognition), a second experiment was run with the same general structure in which birds were fed in both rooms but provided with a different food in each: dog kibble or peanuts. If caching was determined by the fact that birds associate a particular place with a particular food, then one would expect them to cache the food usually found in each room in that same room. Instead, demonstrating their desire for diversity, the birds tended to cache the opposite foods, ensuring that they would have access to both the following morning. As the study's authors note, there is no way of knowing whether semantic or episodic prospection is behind this behavior, but "in either case it shows that these birds must have the capability to plan for a future motivational state [in contrast to simply acting on the current one] over a timescale stretching at least into tomorrow. These results, therefore, challenge the assumption that the ability to anticipate and take action for future needs evolved only in the hominid lineage" (Raby et al. 2007, 920).

Another important study took advantage of an established understanding that birds (and other animals) that are fed only a particular food become "specifically sated," that is, they no longer desire any more of that food. In this study, jays were fed a particular food until sated and then allowed to cache two different kinds of foods. The jays were not able to access their caches until the next morning, however, when they would have just been fed the second type of food for breakfast. The question was whether they would cache the food that they currently preferred or the one that they were currently not interested in but knew that they would prefer at that point in the future when they would actually be recovering and eating the caches. They tended to do the latter. In this study too, then, the results showed that "contrary to the Bischof-Köhler hypothesis, scrub-jays are not bound to their present motivational state. The birds can anticipate and take appropriate action toward the satisfaction of a future need, one that is not currently experienced" (Correia et al. 2007, 859). Other studies have added even more richness to this behavior, exploring the way in which corvids carefully avoid being seen by others when caching their prizes. If they are seen, they often return to the cache at a later time to move it somewhere else. In this way it seems that they anticipate future theft and act to ward it off (Clayton, Bussey, and Dickinson 2003, 690).

Over the years, these kinds of studies have convinced many scientists that at least some animals are anything but "stuck in time." Corvid caching, it seems, embodies a conscious relationship with the future, an imaginative engagement with what might be and a deliberate, mindful effort to bring about some possibilities and not others. But what kind of an imagination is this? Is it going too far to suggest that as an aga tucks an almond into a hole and covers it over, as she carefully checks for potential pilferers, she might enjoy a moment of *hope*? In posing this question I am mindful of the dangers of collapsing all kinds of prospection into a single category that ignores important differences. But I am equally concerned to avoid failing to appreciate similarities and connections. I want to hold open room for the suggestion that we might be in the same general neighborhood here, that these might be what Vinciane Despret (2016b, 176) would call "versions" of the same phenomena or, at the very least, that playing around in this philosophical space might help us see in interesting ways. If, as we have every reason to think is the case, our caching aga is able to imagine a future in which the almond has been stolen by another crow that she saw nearby at the time of caching, if she is motivated to return later to move the nut to a safer place, does it make sense to think that this behavior is *not* grounded in the imagination of, and desire for, a particular future: one in which the almond will still be there when she returns? Can this behavior really be understood in the absence of something like hopefulness? In a world of uncertainty, a world in which food may well be scarce, caching is an effort to bring about the more desirable future in which one is not hungry.

From this perspective, hope is anything but a uniquely human pastime. These provisioning crows help us see that we inhabit a world of multiplying futures, a world thick with both possibilities and with crafty, desirous, living beings working toward their fruition. We simply do not know how many aspects of their lives crows apply this kind of thinking to. What these caching experiments demonstrate is a *capacity* to think in this way. We should by no means assume that caching is the only realm in which crows imagine and work toward a future; in fact, recent research suggests that it is more likely to be a capacity that they can apply much more generally.[14] What else might the crows of Rota hope for? Might they notice and even mourn their own dwindling numbers, the deaths of their kin, the scarcity of a favorite food tree? As they do so, might they imagine an increasingly diminished future? In raising these speculative questions I want to take seriously the possibility that extinction is not an unthinkable thought for all of those nonhuman species that face it; in their own ways, at least some individuals of some species, may see and feel it coming. But, more optimistically, in posing these questions I want to make room for the possibility that these birds might, in their own corvid ways, also imagine other, brighter, days to come.

CULTIVATING HOPES

There is a tendency to view hope as an internal feeling, an optimistic psychological state, taken up by a subject in relation to an external world. From this perspective, hope might be understood, admittedly crudely, as "a feeling of expectation and desire for a certain thing to happen" (*OED* 1989). In this way it can become part of an immaterial space somehow distinct from material reality, distinct from the way the world actually *is*. But such a framing misses much of what hope is and does. As Ben Anderson (2006, 700) insists, drawing on the work of Ernst Bloch, hope and other "anticipating elements" are anything but divorced from reality. Hope is a part of the material order of things, grounded in bodies with specific cognitive and emotional competencies but rooted also in the broader worldly webs of relating that shape the coming-to-matter of specific futures. Hope is an integral part of the ways in which worlds come into being and pass away, of the enactment of what is. In short, hope is a mode of worlding.

Again and again, in conversations in the Mariana Islands, when I asked people if they were hopeful about a future for the aga or about the state of the environment, they did not tell me about how they felt—or rather not simply about how they felt—but also about what they were doing to bring about particular futures, how they were imagining and crafting possibilities. Jeff Quitugua is a case in point. Jeff is a Chamorro man who works as a biologist for the Division of Aquatic and Wildlife Resources on Guam. He is a passionate conservationist, actively engaged in learning everything he can about the history of these islands and of their peoples and other diverse life forms and in sharing that information and enthusiasm as widely as possible. For the last decade of the aga's time on Guam, Jeff was one of the principal biologists monitoring the species, along with his senior colleague Tino Aguon. Now that the species is gone, he misses their presence in the jungle, and each year, for his birthday, he travels to Rota to see aga in the flesh again. He is clearly a man who feels deeply the losses that are unfolding across these islands, yet when I asked Jeff if he was hopeful he replied with an emphatic "yes." And then, immediately, he turned to telling me about his various projects to improve the state of the local environment:

> Even if it's just to go out there and throw out seeds in the forest. Whatever it is, to help contribute. Whether it's picking up trash on the side of the road or in the jungle or at the beach, or whatever. . . . A lot of my work is volunteer in my off-hours. I'm limited with what I can do here in this office. So, from planting trees to rehabilitating forests, those are me getting people out from the community to go out there with me and, you know, start removing invasive species, whether it's weeds, or vines, or whatnot. . . . I'd rather showcase that, and share it with the people, so they can have hope, they can start taking ownership of that. I know several people now who do beach cleanups. They don't have to wait for the annual national coastal cleanup, or for some big group to have a beach cleanup, they just go out there and do stuff now. . . . So, we're not relying solely on the government, we want to take ownership, and that's where hope comes in. Reaching out to the kids and really just sharing a lot of that.

Hope is a project, a labor, in Jeff's account. It is not a vague optimism about the future but something that must be worked on, built, shared with others so that it might take root and grow unpredictably, wildly. Hope is an ongoing effort to *cultivate* the conditions for a better future.

As a mode of worlding, hope is not always grounded in this kind of *deliberate* action; there are many ways in which hopes shape worlds. But it is this kind of active work that interests me most strongly: hope as a work of care for the future. Much more than an anticipation of or a simple desire for a coming good, hope is an effort to care for that possibility in committed, practical, and situated ways. From this perspective, hope does not pre-exist acts of hopefulness. Hope grows and expands as we actively care for the future. It is an iterative, cumulative, self-reinforcing work, an effort to do whatever one can—within the context of the many, always unequal constraints that shape our possibilities—to nurture a better future into being. In this context, hope takes the form of *provisioning.* The *Oxford English Dictionary* tells us that provision involves "foresight; prevision, looking ahead; *esp.* foresight carefully exercised; prudence, care." To "make provision" is to act on this visioning, to make room and take care of and for what might come. As we have seen, this is precisely the kind of hopeful activity that aga are engaged in through their caching behavior. Caching is not wishful thinking, a naïve optimism about what is to come. Rather, it is an *active* work, an effort not just to imagine but to bring about, to, quite literally even if unconventionally, sow the seeds of something else.

Hope as provisioning is a fundamentally worldly, relational work. Jeff's efforts to share his environmental activities with other members of the community—to have them "take ownership" and begin their own projects—is part of allowing hope to grow and spread. It is an act of provisioning. Doing the work of caring for the future fosters hopefulness; that hopefulness in turn animates and guides the labor. More than an empty wish thrown to the wind, hope must here weave its way into and through the world. This worldly materiality is a key part of the difficult work of inhabiting the space between the dual paralyses of an all-consuming despair and a delusional optimism (Haraway 2016, 3–4). Hope as provisioning asks for something other than these two poles, something at once more fragile and more concrete.[15]

This understanding works against Vincent Crapanzano's (2003, 6) suggestion that hope is grounded in passivity. Positioning hope as desire's "passive counterpart," he argues that "Desire is effective. It presupposes human agency. One acts on desire—even if that act is not to act on desire because one has judged it impossible or prefers the desire to its fulfillment. Except where it is used as an equivalent to desire, hope depends on some other agency—a god, fate, chance, an other—for its fulfillment." In contrast, I would suggest that hope is relational and worldly, that is to say, *ecological* (which is not to say that desire is not ecological in its own ways too). While what is hoped for resides beyond the powers of those who hope, this does not mean that their actions are insignificant. Just because someone cannot produce a desired future singlehandedly or at a whim does not mean that they cannot contribute in some way, however small, to its coming to pass. Activity and power are always only possible as and through webs of enabling relationship (Latour 1986). Paying attention to ecologies of hope—to the relationships that enable better futures to be imagined and realized—is vital work. But so too is a careful attention to the breakdown of these relationships, to the undermining of the conditions under which hopes might flourish.

On Rota, the decline of the aga has taken form in and through the dissolution of many relationships, including those between these birds and their almond trees, between aga and a people that long lived with them, and between these people and government conservationists. As these relationships have deteriorated, the aga has declined, and hopes for its future have become harder to sustain. At the same time, declines in tourism and manufacturing in the islands, alongside the highly ambivalent growth of the U.S. military, have shaped and constrained the kinds of futures that might be hoped for and strongly influenced when and in what ways people would like to make use of their lands. Somewhat perversely, the downturn in the local economy has likely helped reduce the tensions over aga in recent years. As one local commented to me: "There's not much land selling or development happening now. So, there's been less issues with the people and the crow. But when the next change brings us to a better economic status, people might want to develop their property or sell it, and I don't know what's going to happen." Paying attention to this context helps explain why things have worked out as they have. But there was nothing "inevitable" about the decisions by some Chamorro people to kill crows and chop down their trees. Even within the context of a long history of colonization and ongoing U.S. intervention on the island, other forms of resistance and cohabitation might have been pursued—and were, in fact, by some. To forget this fact both lets people off the hook too easily and at the same time is a kind of disavowal of their agency—however fraught and constrained—in the shaping of their own lives and landscapes. Other futures were and are still possible. Likewise, government agencies might have done—and might still do—a much better job of actively listening to local people, of getting to "know the people first" (as Stan put it), and in this way both building community involvement and tailoring conservation approaches to their needs.

In the face of this history and genuine ongoing adversity and uncertainty, some locals—both Chamorro and those from the United States—are working to (re)establish key relationships, to cultivate ecologies of hope. The biologists on the island are doing everything they can to bolster links with the local community: from Sarah and Phil's establishment of Luta Bird Conservation, a small NGO with a strong focus on public outreach and education, to Dacia's and Doug's active work to enlist local interns to be part of their conservation projects. Meanwhile, two of the local government land managers, Chamorro men, that I spoke to on Rota expressed a willingness to start planting Pacific almonds and any other trees that might bolster the aga population. This is something that they're already doing for the endangered Mariana flying fox (*Pteropus mariannus*), a species generally loved by local people, in large part because it is a traditional delicacy (albeit one they are no longer legally allowed to eat). But any kind of widespread planting program or broad social acceptance of the aga, as they also quickly pointed out to me, is going to require both funding and changes to existing land-use regulations. Stan, for his part, is already planting endangered trees on his property, but he is a rarity, and the fact that this activity is viewed by many as a little crazy—inviting the presence and scrutiny of conservation officials—tells us that more must be done to open up this possibility on the island.[16]

None of these proposals are easy or straightforward. Any improvement here will invariably require compromises. We do not and cannot hope alone. As such, we must cultivate modes of attentiveness to others' hopes and the broader relationships that nurture them. While corvid caching fascinates me and offers a relatable bridge out into nonhuman prospection, I do not think we should rest here. "Hope" in a multispecies register should be more than hope *for* other species (or even *with* a select few); it must also be an invitation and a challenge to learn to recognize radically other modes of hopefulness, diverse forms of future-orientedness, in which others—human and not—are working toward particular possibilities. Might we also see a kind of hopefulness in an almond tree's production of its own seeds: Will they be taken? Will they be eaten? Will they find soil and germinate? Even absent any (readily recognizable) reflection or anticipation of these possibilities on the part of the tree, might the long evolutionary and developmental processes that have given rise to this investment in, this working toward, a particular future reveal another kind of hopefulness that resides within and productively shapes our world? Such a possibility might be part of a broader effort aimed at, in Val Plumwood's (2002, 179) terms, rediscovering and "restoring the intentionality stripped from the material sphere." "Why," she asked in a posthumously published essay, "can't we see evolution, for example, as a form of experimentation, of testing and learning, like trial and error, a form of wisdom?. . . Dispersing creativity and agency, we can think the possibility of creative, mindful matter" (2009, 125). Again, the goal here should not be to collapse diversity into singular categories but to learn to attend better to rich patterns of continuities and differences. With this in mind, why can't we understand an almond tree as a being engaged in its own distinctive but nonetheless significant acts of hopefulness?

We do not and cannot hope alone. We inhabit a world of multiplying futures, woven through with so many forms of prospection, of provisioning, of preemption. Inhabiting this world well requires us ceaselessly to learn to see in new ways, to appreciate others' unique and distinctive acts of future-making. This chapter is a single step toward this kind of attentiveness, an effort to weave together a few hopeful stories and to explore the relationships that enable or undermine these possibilities. On Rota, in the midst of considerable difficulty, some modest forms of "biocultural hope" (Kirksey 2012) are taking root. Perhaps we can find grounds for optimism in these diverse efforts to (re)establish lines of communication, of relating, of appreciation, and of nourishment. In short, to (re)weave complex ecologies of hope. In exploring these possibilities, this chapter is itself an act of provisioning, an effort to contribute in some small way to the understandings and renewed relationships needed for both people and crows to again imagine and cultivate flourishing worlds. Flitting in and out of my mind's eye as I write has been the image of a single bird: an aga, perhaps even the aga we know as Diseja, still out in the jungle, quietly, diligently caching almonds away for a deeply uncertain future. Haunted by this image, this chapter is a modest act of solidarity with that bird, with that tree, with the many people who make their lives in this place, and with the diverse futures that they and theirs might still have.

Afterword

IN THE WAKE OF TYPHOONS

As I finish work on this book, I have aga on my mind. In late 2018, two massive typhoons slammed into the Mariana Islands. Mangkhut arrived first, in early September. It moved across the island of Rota, its 110-mile-per-hour winds leaving destruction in their wake, including significant damage to houses and infrastructure, from the airport to the power grid (Bautista 2018). The island's crows seem to have weathered the storm quite well, though. While the nests that many of them had already carefully woven together for the breeding season were destroyed, the birds themselves survived. When I emailed Sarah Faegre (introduced in chapter 5) about a week after the typhoon struck to ask after the aga, she told me that she was hopeful that they would re-nest in three pockets of jungle that are less damaged: "In fact," she said, "we just found our first post-storm nesting attempt!" The small group of captive aga also survived unscathed. At the time Mangkhut hit, these young birds were getting ready to be released, the first cohort from the recently established rear-and-release program. Thankfully, Sarah and the other staff relocated the birds to safety before two of the aviaries that they usually occupy were completely destroyed. The release of these birds was delayed by the storm, but a couple of weeks later it went ahead: five feathered bodies were freed into the wider world to bolster the island's dwindling aga population.

FIGURE A.1. Two aga at the release site, Rota. *Source*: Photo by author.

It was only about a month and a half later that supertyphoon Yutu struck. This second typhoon proved to be even more devastating than the first, although in this case the islands of Tinian and Saipan were much harder hit than Rota. Indeed, Yutu was "one of the strongest recorded tropical cyclones to make landfall anywhere on the planet" (D'Angelo 2018). Yutu caused two fatalities and hundreds of injuries across the islands. On the small island of Tinian, "most of the houses were destroyed, even some concrete ones reduced to rubble" (Eugenio 2018), with similarly significant impacts to houses and infrastructure on Saipan (Kelleher 2018). Several weeks after the storm struck, hundreds of people are still living in temporary shelters, and many residents are without sustained power and water (Encinares 2018).

In the hours leading up to Yutu's arrival, it was projected to pass much closer to Rota than it did—a late shift in course that spared the island much of the destruction. But as the storm approached, Sarah and her team made the decision to recapture the recently released aga and keep them in the concrete typhoon shelter for safety. Thankfully, both that building and the aviaries that they had just spent six weeks repairing were spared this time around. Writing a blog post as the typhoon approached, Sarah noted that earlier that day she had "watched three of the released Aga flirting with small gusts of wind" and thought to herself that this storm "will be the first of many, during their long lives on this little island in 'typhoon alley'" (Faegre 2018).

Despite the ferocity of Mangkhut and Yutu, the aga seem to have made it through both storms relatively unscathed, with a lot of help from people and more than a little luck. But the effects of events like these are not always immediately obvious. Emailing with Earl Campbell in the aftermath of these typhoons, I was surprised to learn that one of the key concerns that these storms have raised for ongoing conservation work in the islands relates to the movement of brown treesnakes (the species that extirpated the aga on Guam). Earl is an invasive species program manager for the U.S. Geological Survey and frequently works in the Pacific. He explained to me that keeping other islands in the Marianas free of this predator involves significant ongoing monitoring: "On Guam, there is a federal quarantine program with fifty-six staff and thirteen dogs which works to keep snakes out of the transportation network." Smaller programs perform the same function on islands in the CNMI. When a big storm hits, these processes can become disrupted, while the movement of aid and other resources between islands can be significantly scaled up. In short, Earl explained, "storms pose heightened risk for snake transport." This is a high-stakes situation. Were the brown treesnake to become established on Rota, it is highly unlikely that the extinction of the aga could be prevented (alongside the extinction or extirpation of many of the island's other birds and reptiles).

Climate change enters the story of the aga here as yet one more possible factor working against the species' long-term survival. While the Mariana Islands are no strangers to extreme weather—as is clear from Sarah's reference to "typhoon alley"—in this age of climate change we can only expect to see an increase in the frequency and intensity of these kinds of events in this part of the Pacific (Leong et al. 2014). It isn't really clear what this might mean for the aga. If frequent typhoons and storms kill birds or prevent them from successfully fledging chicks, this could be significant for this already small population. But perhaps it is the indirect effects of climate change we should be more worried about. The movements of brown treesnakes are one aspect of this, but so too is the capacity for climate change to place more pressure on the relationship between local people and aga, for example by undermining the livelihood possibilities associated with agriculture and tourism. In short, it seems increasingly likely that on Rota—as is the case all over the world, but in different ways—climate change will be a major factor in shaping and curtailing the kinds of futures that might be hoped for, the kinds of futures that might be *possible* for everyone.

Perhaps it should be hard to tell stories about crows in times like these. In the midst of so much human suffering, of mounting challenges, of long histories of injustice and resistance to it by local people, what does it mean to hold the feathery forms of aga and other corvids at the center of our narratives? What does the well-being of an individual bird, or perhaps even the continuation of their species, really matter in the wake of increasingly destructive typhoons? Indeed, the people of the Marianas are themselves in need of powerful stories that might reach distant audiences: even as they suffered through these storms and live on in their aftermath—some of them still without reliable power and water—much of the United States seems not to have noticed. Yutu was, as Anita Hofschneider (2018) put it, "The super typhoon American media forgot," and, it should be noted, there was even less coverage of Mangkhut's effects on Rota. U.S. president Donald Trump didn't even bother to tweet about Yutu, a silence that "stands in stark contrast to his public response during hurricanes Harvey, Irma, Florence and Michael, which struck the U.S. mainland" during his presidency (D'Angelo 2018). But, like it or not, the fates of both the people and the birds of the Marianas, while far from being exhaustively shaped by the United States, are tied in myriad consequential ways to this distant power, to its military, its conservation programs, and its greenhouse-gas emissions.

This book is a response to precisely these kinds of challenges, a response to the profound difficulties and uncertainties that seem to characterize our time. *The Wake of Crows* is grounded in the conviction that we all have multiple, overlapping accountabilities to others: to diverse people, to crows, to the countless others that comprise the wider webs of living and dying that hold us all in shared worlds. But more than the simple acknowledgment of multiple accountabilities, this book arises out of the understanding that none of these accountabilities can really be taken up *well* in isolation: cultivating flourishing worlds requires modes of attention tuned to thick, multiplicitous, biocultural relationalities, with all of their complex histories and possible futures. It requires that we do not turn away from those processes of wreckage that are shaping so much of our contemporary world (Rose 2006) but instead turn toward them, that we learn to attend to the shifting, profoundly consequential ways in which we are all caught up with and at stake in one another—that we turn into the wake, with ever more care, humility, and appreciation for what remains and might still endure.

NOTES

INTRODUCTION: MAKING WORLDS WITH CROWS

1. Abiotic entities and processes embody and enact their own modes of lively responsiveness (TallBear 2017, Reinert 2016, van Dooren and Rose 2016).
2. These aspects of corvid behavior are discussed further in chapter 4.
3. While there are a great many exceptions to this situation around the world, those crows that are not exceedingly rare are quite frequently viewed as overabundant, at least by their critics (often with reference to notions of "natural balance" and "fit," discussed further in chapters 1, 3, and 4).
4. Zooming out a little, the genus *Corvus* is part of the family Corvidae, which also includes jays, (Eurasian) magpies, choughs, and others. These birds are sometimes all referred to as "crows," but I have reserved the term for the genus *Corvus* and used the alternative term "corvid" when referring to this larger family. Members of this family, especially the jays, appear at various stages in this book, in particular with reference to some areas of cognitive research where there is a more detailed understanding of jay biology than there is of any of the crows. As discussed in more detail later in the book, as close relatives of the crows there is usually every reason to suppose that these capacities are widely shared and relatively similar among the corvids.
5. Discussed further in "Cooperating."
6. In focusing on crows in this way rather than on a more diverse set of species, this book works to tell stories that range out into diverse contextual terrains—exploring processes of colonization, urbanization, conservation, globalization, trade, development, waste, and more—without the nonhumans disappearing into the background. My hope is that these similarities among crows allows the diverse chapters—with their wide-ranging geographical, cultural, and thematic concerns—to fold into one another in ways that enliven and thicken both the individual stories

that I am telling and the presence of crows in the book, providing a fuller sense of who these birds are and how they inhabit and ultimately shape their/our worlds.

7. See Swift and Marzluff (2018, 2015). Also discussed in van Dooren (2014, 125–44).
8. Here, and elsewhere in this book, I am drawing on Karen Barad's (2007) notion of "intra-action" as a form of relating in which entities remake one another in/through their engagements.
9. Some key sources on ethics of/as worlding include Haraway (2008); Barad (2007); Stengers (2010); Puig de la Bellacasa (2017); Alaimo (2016); Shotwell (2016); Åsberg (2013); Ginn, Beisel, and Barua (2014); Neimanis (2016); Giraud et al. (2018); Hollin et al. (2017); Grebowicz and Merrick (2013); Münster (2014); Pacini-Ketchabaw and Nxumalo (2015); Instone and Taylor (2015); Reinert (2016); Rose (2007); van Dooren and Rose (2016); and van Dooren (2014).
10. Astrida Neimanis has productively challenged me to consider the importance of specifically feminist theory in the body of work taking form here. It is undoubtedly the case that much, indeed the vast majority, of the key work on this kind of ethics is emerging out of or drawing substantially on feminist thought. In particular, this seems to be the case because of a long tradition of feminist ethical thought that emphasizes the value of multiplicity and diversity, the need to engage with situated complexity, the entanglement of the material and the semiotic, the interwoven and mutually reinforcing nature of modes of oppression, and the material and practical nature of ethics (not in addition to but as a mode of inquiry). In different ways, each of these dimensions of feminist approaches to ethics is central to the stories told in this book.
11. Here, ethics cannot take the form of an adjudication over how one ought to act in an already given and preexisting reality. Instead, it becomes inescapably entangled with the world's *becoming* and so with questions of epistemology, ontology, and politics, opening up the need for what Karen Barad (2007, 185) has called an "ethico-onto-epistem-ology." This ethics is an ongoing inquiry into the kinds of worlds that we imagine and help bring about: What do they mean and how do they matter for the many beings that co-inhabit/co-constitute them? It is also a creative exploration, and indeed practice, of imagining and bringing about better possibilities—while, of course, attending to the question of what "better" means and for whom. In short, in a context in which we are always making worlds with others, ethics is a process of "taking account" and "being responsible" for our involvement in these processes (Barad 2007, 185, 391–96).

    But things are not quite this simple. If our worlds are constantly in the making, so are we. From such a perspective the human subject is radically decentered, no longer the sole locus of (ethical) agency or of ethical concern. Ethics has often been grounded in "universalizing conceptions of the ethical subject as an autonomous, rational, and defined 'self'" (Puig de la Bellacasa 2017, 137): a being who freely authors its own actions and so can be held accountable for them. From a worlding perspective such a subject does not exist. Subjects are all continually "in the making" inside larger patterns of worlding, and so both agency and responsibility must be understood in more partial, relational forms. But even in such a context—in which living beings are far from "individual" and "autonomous" in the traditional senses of these terms (Neimanis 2016, Alaimo 2016)—we might still insist that many of us are, nonetheless, beings who are in important ways capable of both agency and responsibility. Ethics is often framed here in terms of "response-ability" (Haraway 2016, Stengers 2014); whether and how we respond to others is always contextual, being made newly (im)possible, and so ethics becomes in large part a process of cultivating the *capacity* to see and to respond well. As such, ethics cannot simply be something that one *does*, a mode of inquiry and action; it is also something that one *undergoes*, is transformed by, becomes otherwise through. In such a context, ethics is in the details: the particularities of the who, how, and with what consequences of worlds in the making.

In connecting questions of responsibility to those of how one is *able* to respond, this framing places new emphasis on the broader contexts through which one is rendered capable of response in the particular ways that one is, from evolutionary history, which has provided us with particular cognitive and emotional competencies, through to our own life histories, education, freedom, and capacity to understand and to act. In doing so, we move beyond simplistic and singular framings of "human" responsibility within a passive and inert more-than-human world. Each of us, human and not, is positioned differently—and as such we are each differently response-able. As Deborah Bird Rose and I (2016, 90) have argued elsewhere: "there is no singular 'responsible' course of action; there is only the constantly shifting capacity to respond to another. What counts as good, perhaps ethical, response is always context specific and relational. It is always being rearticulated, re-imagined, and made possible in new ways, inside ongoing processes of call and response and the worlds that they produce. Here, responsibility is [in large part] about developing the openness and the sensitivities necessary to be curious, to understand and respond in ways that are never perfect, never innocent, never final, and yet always required."

In some renderings of worlding ethics, any kind of deliberation and intentional action by subjects is thoroughly backgrounded in favor of a much more expansive sense of ethics as mattering. For example, Karen Barad (2010, 265) has argued that "crucially, there is no getting away from ethics on this account of mattering. Ethics is an integral part of the diffraction (ongoing differentiating) patterns of worlding, not a superimposing of human values onto the ontology of the world (as if 'fact' and 'value' were radically other). The very nature of matter entails an exposure to the Other. Responsibility is not an obligation that the subject chooses but rather an incarnate relation that precedes the intentionality of consciousness. Responsibility is not a calculation to be performed. It is a relation always already integral to the world's ongoing intra-active becoming and not-becoming. It is an iterative (re) opening up to, an enabling of responsiveness."

While I want to hold open this expansive sense of ethics, I also want to insist that we pay attention to the diverse ways in which different beings inhabit these spaces of mattering. As noted earlier, some beings, some subjects, do engage in varieties of intentional world making grounded in deliberate efforts to explore and enact various goods (however relational intentionality, deliberation, and action must be). In some cases, crows and other nonhumans might be among their number (a related discussion of diverse forms of agency is offered in chapter 4). Attending to these specificities matters, multiplying what ethics might mean, a process that need not involve restricting the ethical to the domain of rational, autonomous subjects (whatever these might be).

As such, in this book I frequently connect ethics to notions of working, crafting, cultivating, and making. All of these broad categories of activity take place within but are also constitutive of relationships and worlds. In my use of these terms, I aim to reference possibilities for more or less intentional practices, purposeful involvements, in the bringing about of worlds. Worlds are always made from the inside, with and against specific others. As Dimitris Papadopoulos (2014) notes, making is not about autonomous production: "Making starts from what is there. Intensive recycling. Immediate caring. . . . We make as we coexist in ecological spaces." This is what Donna Haraway (2016, 58) has called *sympoiesis*, a term that "means 'making-with.' Nothing makes itself; nothing is really autopoietic or self-organizing."

12. On this aspect of Haraway's approach to theorizing, also see Grebowicz and Merrick (2013, 14).
13. The focus here, as throughout this book, is on the biotic. But multispecies studies need not exclude other entities and processes—and certainly shouldn't do so a priori.
14. Clare Palmer (2010) compellingly argues that one of the key problems with many of the dominant approaches to animal ethics is that they tend to consider animals in the abstract but also in isolation—focusing on capacities without enough attention to the specific contexts that must always influence what various humans owe to animals, no matter how similar they might be in terms of, for example, their sentience.
15. I take the terminology of universal consideration from Thomas Birch (1993) but am equally influenced by the general approaches in this area developed by Val Plumwood (2002, 167); see also Rose (2013b), Cheney and Weston (1999), Warkentin (2010), and others. In my terms, it is a recognition of and respect for the diversity and multiplicity of forms, living and otherwise, that constitute worlds, such that these others *need not justify their value or existence in order to be deemed to be ethically relevant.* This understanding is in deliberate contrast with most approaches to animal ethics, which are premised on "extensionist" logics in which ethical considerability is extended to others on the basis of their possession of criteria that we value and recognize in ourselves, like sentience or rationality. As Plumwood (2002, 167) has noted: "Such a schema based on sameness to the human treats earth others as of value just to the extent that they resemble the human as hegemonic centre, rather than as an independent centre with potential needs, excellences and claims to flourish of their own." (Of course, as Plumwood was aware, these invocations of "the human" are themselves always grounded in forms of privileging and exclusion. See chapter 3 for a fuller discussion of her thought in this area.) This understanding also supports the broader practice of ethical attending I take up in this book. Beginning from a fundamental respect for others on their own terms, ethics necessarily becomes a highly contextual affair, a work of paying attention to "other's needs, ends, directions, or meaning" (Plumwood 1993, 138) as they take form within particular, concrete situations. As Warkentin (2010) notes, far from being a generalized ethical "system" that produces "one-size-fits-all" solutions, this kind of ethics is grounded in a work of attentiveness or etiquette (see also Cheney and

Weston 1999). But this is only a starting point; an opening or invitation to attend to others must in many cases ultimately still involve the working through of whose possibilities will thrive and whose will wither. In starting with consideration and attentiveness, however, we begin from the supposition that we do not yet know what others might be capable of, why they might matter, and how we might craft possibilities together (Despret 2008b). In this way, we give others the greatest possible chances, and we hold open room for ourselves to be transformed: to learn to see, and so to be, differently (Cheney and Weston 1999).

16. As Eva Giraud and colleagues note: "There is a danger . . . if a non-anthropocentric ethics begins and ends by recognising that the world is constituted by and through entangled, more-than-human relations. It is important, we argue, to engage in more sustained attempts to identify and meet the ethical obligations posed by particular relations, which go beyond the recognition that no course of action is innocent" (Giraud et al. 2018, 74).
17. See, for example, the "Living Lexicon for the Environmental Humanities," published in an ongoing manner in the journal *Environmental Humanities*, and the "Lexicon for an Anthropocene yet Unseen," published by *Cultural Anthropology*, and the recent *Keywords for Environmental Studies* (Adamson, Gleason, and Pellow 2016).

EXPERIMENTING

1. On corvid recognition of human faces and cars, see chapter 4.
2. The ethics of interspecies experimentation is a major theme of chapters 1 and 4.
3. On the need to think "the human" with specificity, perhaps more than ever in these Anthropocenic times, see chapter 3.
4. This situation and the possibilities for multispecies experimentation are discussed in more detail in chapter 1.
5. Corvid neophobia is discussed further in chapters 1 and 4.

1. INTERJECTING CROWS: ENACTING MULTISPECIES COMMUNITIES

1. For recent research on the ways in which American crows interact with the dead of their own kind, see Swift and Marzluff (2018, 2015). I have discussed corvid grief and funerals in more detail in van Dooren (2014, chap. 5).
2. The use of these kinds of alternative terminologies for killing is discussed in chapter 4.
3. Unless published work is cited, references to Darryl Jones's ideas and opinions draw on interviews conducted by the author in person or over the phone on multiple occasions between early 2015 and late 2017.
4. My own reading of newspaper articles on crows in the Brisbane region (beginning from the 1850s) supports but cannot conclusively confirm this arrival date. The earliest detailed studies of Australian crows and ravens were conducted by Ian Rowley (1973) and focused predominantly on rural areas.

5. Siobhain Ryan, "Locals Call for Culling of Crows," *Courier Mail*, March 23, 2002.
6. Frances Whiting, "Chorus of Crow Haters," *Sunday Mail*, May 31, 1998; Siobhain Ryan, "Cawing Prompts Call for Murder of Crows," *Courier Mail*, October 14, 1999; Ryan,, "Locals Call for Culling of Crows"; "Crow Haters Seek Help," *Courier Mail*, June 26, 2004.
7. "Crow Haters Seek Help."
8. Here and elsewhere in this chapter I have drawn on a series of interviews conducted with employees of various government agencies, wildlife-management companies, the Royal Society for the Prevention of Cruelty to Animals (RSPCA), and a range of wildlife carers, bird watchers, and others in the Brisbane area between early 2014 and late 2016.
9. It is worth noting that the particular characteristics of subtropical Brisbane weather mean that houses are often kept open to the breezes, and so the noises, especially in the warmer months. While drafting this chapter I lived in Germany for a period of three months, directly alongside a roost that usually accommodated roughly three hundred carrion crows (*Corvus corone*) but on occasion as many as a thousand. While the birds were very noisy, it was immediately apparent what a difference a closed (double-glazed) window made—in addition to a well-insulated building. From this perspective, both the Queensland weather and Australia's generally lax building standards are likely working against harmonious human-crow relations in Brisbane.
10. In Brisbane, as in much of eastern Australia, it seems that the growing abundance of the native noisy miner (*Manorina melanocephala*)—as a result of diverse forms of anthropogenic habitat disturbance—is a much more significant issue for avian diversity (Kath, Maron, and Dunn 2009; Maron et al. 2013).
11. See the discussion of watching crows and crows watching back in chapter 4.
12. See chapter 4 for a discussion of the complicated dynamics of abundance and "doing well" as a species. See chapters 2 and 5 for discussions of struggling crow species.
13. Deborah Bird Rose and I have explored these themes elsewhere: van Dooren and Rose (2016) and Rose and van Dooren (2016).
14. Darryl Jones cited in Watson (2017).
15. The common name "magpie" is widely used for a range of unrelated black and white birds. Unlike the Eurasian magpie (*Pica pica*) and some other magpies, the Australian magpie (*Cracticus tibicen*) is not a corvid. For a detailed discussion of this situation, see the entry of the wonderful *Corvid Blog* by Jennifer Campbell-Smith, http://coyot.es/thecorvidblog/2016/03/07/australian-magpies-are-not-corvids.
16. Clemens Driessen (2014, 91) has likewise advocated for "an inquisitive, experimental, ongoing politics of everyday animal encounters."
17. I am drawing here, in particular, on conversations with Darryl, who took an active interest in these community attitudes during this period.
18. Instead of seeing a community as an organism, as Clements proposed, a great deal of recent work in the biological and social sciences has emphasized the way in which organisms are actually communities in a range of consequential ways (Schneider and Winslow 2014; Gilbert, Sapp, and Tauber 2012; McFall-Ngai et al. 2013; Lorimer 2016).

19. Of course, questions remain about *which people* are asked to do the adapting in these kinds of scenarios and about the need for ongoing critical intervention into these dynamics (discussed in more detail in chapter 3).
20. Naturalist History Society, "Excursion to Stradbroke," *Queenslander* (Brisbane), June 11, 1892.
21. W. Lyon, "The Crow and the Flying-Fox," letter to the editor, *Queensland Times* (Ipswich), December 18, 1912.
22. "In Defence of Birds," *Telegraph* (Brisbane), October 24, 1919. I will return to the topic of introduced species further on.
23. "Concerning the Crow," *Northern Miner* (Charters Towers), August 18, 1913 (reprinted from the *Sydney Morning Herald*).
24. Thank you to Margaret Cook for conducting this archival research.
25. For discussion of newly arrived—invasive and introduced—species in Australia, see Low (2002), van Dooren (2011), and Head (2011).
26. Ryan, "Cawing Prompts Call for Murder of Crows."
27. In particular, see McLauchlan's (2018) discussions of "care smuggling" (chap. 3) and "well-aligned cares" (chap. 5). On the necessarily compromised nature of caring also see Murphy (2015), Puig de la Bellacasa (2017), and van Dooren (2014).
28. Ryan, "Locals Call for Culling of Crows."
29. Nature frequently stands in for "the good" in this way and the unnatural for the bad. In classical thought, as Daniel B. Botkin (1990, 8) notes, the natural balance was "that constant condition which was desirable *and good*" (emphasis added).
30. Of course, these challenges are not at all neatly separable within the complex contexts of communities and worlds-in-the-making.
31. On, or rather against, an ethics of purity, see Shotwell (2016).
32. In thinking with Haraway in this way about the constitution of community as an ethicopolitical work, I might be accused of stretching her work on "situated knowledges" in unreasonable ways. In contrast to such a position I would insist, as Grebowicz and Merrick (2013) do, that these kinds of commitments are very much at the heart of Haraway's work here. Indeed, as Haraway (1991, 579) notes of her own project: "in traditional philosophical categories, the issue is ethics and politics perhaps more than epistemology."
33. Connolly (2005, 4) describes this as an orientation in which an individual must simultaneously hold their own "faith creed or philosophy" *and* "exercise presumptive receptivity towards others when drawing [it] into the public realm." I am not convinced that "bicameral" is quite the right term here—something more multiplicitous seems necessary—nor am I sure about how to make the cut between the "public realm" and something else, but Connolly's general point is instructive.
34. Personal correspondence with Gisela Kaplan, May 24, 2017.
35. These modes of communication have the potential to become "diplomatic proposals" in the sense discussed in chapter 4.
36. In important ways, these are precisely the kinds of approaches that I have called for in other contexts. See, for example, chapter 3.
37. On human-animal friendships, see Lestel (2014a) and Kirksey et al. (2018). Of course, this strategy is also limited by the fact that it is likely only possible for

people who spend considerable time within a given magpie (or crow) territory, not those who are just passing through.

38. Interrogating these larger dynamics is a more central component of the remaining chapters of this book, in particular chapters 2, 3, and 5.
39. In raising the admittedly speculative possibility that Brisbane's crows may also be engaged in their own kind of holding open of room for others, I want to insist on the need for a loose and uncertain understanding of what this might mean. Rather than firm and rigidly defined definitions of situated pluralism and the work of making community well, I would prefer to remain open to the many different, unexpected forms this might take, using these terms to explore spaces of shared life with others that are attentively making room, connecting, and proposing. "Targeted helping" is a highly sophisticated cognitive and empathetic capacity (de Waal 2008), and many animal species likely are not interacting with others in this way. Might they have their own forms of attending to and making room for others?

STEALING

1. Alongside the cited materials, this account draws heavily on an interview with Nicola Clayton, conducted by the author at Cambridge University on May 19, 2017.
2. For a fuller discussion of caching behavior, see chapter 5.
3. These experiments are discussed in more depth in chapter 4.
4. Theory of mind is discussed in more detail in chapter 4 and is also raised in passing in "Gifting."

2. SPECTRAL CROWS: CONSERVATION AND THE WORK OF INHERITANCE

1. Interview with Donna Ball conducted via phone on October 17, 2018.
2. The date of Polynesian settlement of the Hawaiian Islands remains a topic of ongoing research. For a review and an argument that this event likely took place between 1000 and 1200 CE, see Kirch (2011).
3. For a longer discussion of the coevolution of calls and their reception, see van Dooren and Despret (2018). On botanical messages to pollinators, see Rose (2012a).
4. Jeff Burgett (agent with the U.S. Fish and Wildlife Service), interview with the author, Hilo, Hawai'i, December 19, 2011.
5. As Paul Griffiths and Russell Gray (2001, 196) put it, the concept of inheritance ought to be applied "to any resource that is reliably present in successive generations, and is part of the explanation of why each generation resembles the last."
6. In Deborah Bird Rose's (2012b) terms, life is a product of both sequential and synchronous relationships and inheritances.
7. This understanding draws on an extended discussion in van Dooren (2014).
8. Private land is one of the key obstacles here. In some cases, privately owned lands are being closed off to hunters (perhaps because of insurance concerns or landowners' bad past experiences with hunters). In other cases, public land where

people might hunt is inaccessible because the owners of private properties surrounding it—often remnants of large plantations or ranches—restrict direct or open access to it. In addition, it should be noted that relatively little state land is actually utilized solely (or even primarily) for conservation purposes (Lisa Hadway, interview with the author, January 25, 2013; Hadway is the manager of the state government's Natural Area Reserves System, Division of Forestry and Wildlife, Department of Land and Natural Resources). At present, the Division of Forestry and Wildlife (DOFAW) provides 600,000 acres of public hunting land on the island of Hawaiʻi. Of this land, "only about 4 percent is currently fenced with hooved animal populations effectively controlled [a requirement for effective conservation]. Under the most ambitious current plans for fencing and ungulate removal over the next decade, about 17 percent of DOFAW lands on the island would be affected, most of which would occur on Mauna Kea" (Geometrician Associates 2012, 86).

9. Anonymous interviewees. Unless otherwise noted, these interviews were conducted by the author with biologists, managers, hunters (including both Native Hawaiians and other locals in all three categories), in January and February 2013 on the islands of Hawaiʻi and Oʻahu. In most cases, I have identified participants by name; in a few cases, where more appropriate, I have referenced them anonymously.
10. I accept J. Kehaulani Kauanui's (2009) argument about the appropriateness of the term "colonization" to describe the social and political dynamics of Hawaiian life after what was technically an "occupation" of the internationally recognized sovereign nation of Hawaiʻi. See also Silva (2004).
11. Of course, around the world cultural and linguistic diversity also often relies on biodiversity, and vice versa (Maffi 2004; Martin, Mincyte, and Münster 2012).
12. This comment was either made directly to me or presented by others as a claim commonly made, in several anonymous interviews conducted in January 2013. Similar comments can be found posted to the hunting forum Hawaii Sportsman, http://hawaiisportsman.forumotion.com/t5382-big-island-video-news-hunters.
13. Also see Gon (unpublished) and Maly, Pang, and Burrows (2007).
14. The Hawaiian term *haole* is widely used in Hawaiʻi to refer to people who are not of Native Hawaiian descent, often specifically Caucasians.
15. The Great Mahele was a period of land redistribution—initiated by the king and the parliament of Hawaiʻi—that "converted" traditional customary rights in lands into private property in the mid-nineteenth century, in the lead-up to U.S. occupation (Banner 2007, Silva 2004).
16. *Lāhui* is a multiplicitous Hawaiian term that means nation, gathering, people, and tribe. As in this case, it sometimes used to evoke the tight relationship between people and nation.
17. "Blue Mountain Traila," comment posted to Hawaii Sportsman, June 6, 2012, http://hawaiisportsman.forumotion.com/t5382-big-island-video-news-hunters.
18. "Shrek," comment posted to Hawaii Sportsman, June 9, 2012, http://hawaiisportsman.forumotion.com/t5382p15-big-island-video-news-hunters. There does seem to be something to these arguments. Interviews that I conducted with conservationists, alongside their own public submissions during the community consultation

process for the Kaʻū Forest Reserve Management Plan, make clear that most of them see protecting only 20 percent of the area as, in effect, sacrificing 80 percent. Many of them would like to see a lot more of the area fenced and ungulates removed. It is unclear exactly where the state government stands on this, especially in the long term. Its position seems usually to involve some sort of middle ground that leaves both sides equally unhappy.

19. Shalan Crysdale (an ecologist with the Nature Conservancy), interview with the author, Naʻalehu, Kaʻū, February 7, 2013.
20. For a fascinating discussion of contemporary Hawaiian scholarship that engages with the *Kumulipo*, see McDougall (2016). Although there are a variety of cosmogonic stories in Hawaiʻi, Katrina-Ann R. Kapāʻanaokalāokeola Nākoa Oliveira (2014) notes that a common element shared by many of them is "the genealogical relationships between the land, humankind, and the gods."
21. For a fascinating and in-depth discussion of the proposal to build a series of new telescopes on the sacred volcano Mauna Kea, see Terrell (2017). On this topic also see Goodyear-Kaʻōpua (2017) and Kuwada (2015).
22. Derrida seems to be thinking here about "Life" in a narrower sense than I am, with quite a tight focus on tradition, culture, and language (in human and, in particular, philosophical contexts).
23. Derrida's primary concern in his discussion of responsibility and inheritance is political conservatism and those modes of inheritance that uncritically take up and perpetuate the past. In this context, responsibility emerges as a radical questioning of what is to be retained and what lost or transformed. In Derrida's terms, it is only through "reinterpretation, critique, displacement, that is, an active intervention . . . that a transformation worthy of the name might take place; so that something might happen, an event, *some* history, an unforeseeable future-to come" (Derrida, Roudinesco, and Fort 2004, 4). The basic point here is simple and powerful. Inheritance that is mere repetition closes off the future, or rather, it closes off the possibility of anything genuinely different and maybe, just maybe, better. Thanks to Rosalyn Diprose for her reading of Derrida and for being willing to discuss responsibility and inheritance with me. For a fuller discussion of Derrida's notion of a responsibility "worthy of the name," see Diprose (2006).
24. I have no particular authority to speak on this matter in Hawaiʻi. I am drawn by a genuine concern for the future of these forests and all their inhabitants to attempt to weave my way through these difficult topics, to arrive at some sense of "where to from here." Ultimately, however, I do not intend to argue for the "right to an opinion" on this topic. This chapter is written in large part against the proposition that some people might be shut out of conversations that aim to imagine what responsibility and justice might look like in multispecies and multicultural worlds solely on the basis of the kinds of inheritance that they bring with them: that they don't have the right kinds of history and so knowledges or experiences. Furthermore, from my perspective, there is an ethical demand issued on all sentient creatures to respond when they are witness to suffering, violence, death, and extinction.
25. On the complex intersections between sovereignty and conservation, see Mawyer and Jacka (2018).
26. On the multiplicity of meaning (and worlds) in Hawaiian thought on *kaona*, see Arista (2010). Also see McDougall (2016).

COOPERATING

1. On this last point, also see "Stealing."
2. The significance of cheating for future cooperation was not measured in the rook experiment.
3. A fascinating story told to me by Trixie Benbrook. See chapter 1 for a fuller discussion of Trixie and her work.
4. To borrow a somewhat apt piece of terminology developed for chickens, one that also reminds us of the particular effects of captivity on the *production* of the dominance regimes in those populations that are often most closely studied, i.e., captive ones (Rowell 1974, Despret 2008a).
5. When I asked Nicola Clayton about this, she noted that she suspects that there might be a more general sense among a monogamous mating pair that they share in a way that is reciprocal.

3. UNWELCOME CROWS: HOSPITALITY IN THE ANTHROPOCENE

1. Interview with Colin Ryall, conducted by the author in August 2014. Ryall is a biologist at Kingston University London who has studied house crows all over the world and provided input into the Hoek van Holland management effort. All subsequent references to Ryall (other than to his published work) draw on this interview.
2. See Ruling 201306221/2/A3 of the Common room—Appeals—Flora and fauna (Thursday August 8, 2013).
3. Derrida's broader body of work on hospitality is much more complex and interesting than this. This work is not *required* to understand the key points I want to make in this chapter, but it does enable a few interesting parenthetical observations (which will be made in later footnotes).

   In this work Derrida argues that all hospitality is situated within the contingency, the conditionality, of a world of lived relations that necessarily delimits the space of welcome: who and what are we able to welcome when? But his work aims to push us out beyond the comfortable confines of any particular practice of welcoming in a constant effort to revise, and perhaps better, our hospitality. For Derrida, this effort is grounded in the notion of an "unconditional" or "unlimited" hospitality in which the host gives completely, without limit or condition to whoever arrives. In his words: "Let us say yes to *who or what turns up*, before any determination, before any anticipation, before any *identification*" (2000b, 77). Derrida's point here is not that we ought to adopt this unlimited hospitality in our dealings with others; he notes that it can also produce terrible outcomes (2000b, 151–155). Rather, he argues that this form of hospitality is needed as a constant corrective, as a demand that pushes us to work toward a hospitality "worthy of the name" (Derrida 1999, 2005; Patton 2004). (On the notion of being "worthy of the name," see Oliver [2009, 123]). Ultimately, as Derrida notes, this unlimited hospitality is an *impossible* hospitality. To "achieve" universal hospitality would, at the very same time, undo the possibility of hospitality: to throw one's "home" open to others in this most radical of ways would be to undo one's status as host and lose the material means necessary to ensure that one can continue to host others in this place-time. In addition, we cannot hope to welcome everyone. Derrida knows this; his approach to hospitality aims to hold us between this brute fact and an impossible demand. Here, we are forced to make *decisions* about who will be excluded, why, and how (Lawlor 2007, 112). The *decision* is, of course, central to Derrida's effort to shift ethics beyond the space of the calculable. It is in the *tension* between existing forms of hospitality and this unlimited form that change for the better might happen (Derrida 2005, 6).

But Derrida offers very little by way of situated reflection on the *terms* of our hospitality, on what might constitute *better* forms of hospitality. This is the case, in large part, because of his desire to hold open the future, not to think that we know, or can know, all of what hospitality might be. As Matei Candea (2012, 546) notes, Derrida's vision of hospitality is largely "unmoored" from "the concrete objects and forms which, in practice, allow people to decide where welcome ends and trespass begins." I share with Derrida an interest in more open approaches to the future, approaches grounded in not trying to control, shut down, or even fully predict what is possible (see chapter 1). Contra Derrida, however, I see the greatest possibility for this openness in a move outside of the space of hospitality altogether, a topic that I will take up in the final section of this chapter.

4. There is an extensive literature, across both the human and natural sciences, exploring the ways in which notions of belonging and fit get worked out in relation to so-called introduced species. Some representative articles include Trigger et al. (2008), Robbins (2004), and Davis et al. (2011). This literature has frequently explored similarities between rhetoric and practice in this domain and in that of the human immigrant, who is also often rendered "out of place." There are, no doubt, questions to be asked here—perhaps especially in the context of the rise of the European Right—about the parallels between various acts of (un)welcoming and the forms of national and regional purity and security that they imagine and enact. In this chapter, however, I would like to hold the focus on these crows and as such have largely set this issue aside.
5. On the world-forming consequences of the histories we imagine and tell, see chapter 2.
6. There is another interesting connection to Derrida's work on hospitality here, grounded as it is in a distinctive orientation toward the future. For Derrida, hospitality is "a matter of opening ourselves to meanings to come, meanings that we cannot anticipate" (Oliver 2009, 123). In his own words: "What we call hospitality maintains an essential relation with the opening of what is called to come [*à venir*]. When we say that 'We do not yet know what hospitality is,' we also imply that we do not yet know who or what will come" (Derrida 2000a, 11). The house crows of Hoek van Holland remind us that the future can work on our hospitality in a variety of ways; it can shut down just as it potentially opens up possibilities for welcoming.
7. Short excerpt from the audio that accompanied the animation. Author's transcription.
8. See, for example, the papers in "Imagining Anew: Challenges of Representing the Anthropocene," special issue, *Environmental Humanities* 5 (2014), ed. Greg Garrard, Gary Handwerk, and Sabine Wilke.

9. A point also noted by Szerszynski (2012).
10. From this perspective the Anthropocene can be understood as the latest example of a long line of efforts to name the growing destructive impact of humanity as something distinct, as ushering in a new time. As Clark (2014, 22) notes, "Speculations about a novel human-induced geological period go back at least as far as Italian geologist Antonio Stoppani's coining of the term the 'Anthropozoic era' in the 1870s."
11. Is it *only* human agency responsible for shaping the world now? Of course not. To some extent other-than-human agency is backgrounded and humanity set apart from the rest of the world in the choice to focus on the anthropogenic that is inherent in the label "Anthropocene." But this need not be the case. Instead, we might take up the melding of human and geologic in this term as a provocation to continually question *how* worlds are constituted in entanglements that are always multiagentic, always more-than-human. Clark (2014, 26) notes that "researchers in the social sciences and humanities still tend to treat natural and social agency as sliding points on a linear scale, analogous to a tug of war in which one side gains as the other loses." It is, in part, this simplistic framing that allows the common slippage, especially in popular presentations of the Anthropocene, as a period in which humans have *taken over* the shaping of the earth and are completely in control or at least in which any caution and nuance in the expression of this agency is absent.
12. Eileen Crist (2013a, 48) has pointed to a closely related pattern of self-reinforcing entitlement, noting that an attitude of what she calls "human supremacy" has both enabled the destruction of the Earth, while at the same time this "seemingly triumphant domination of Nature has cast an aura of truth over the belief of human supremacy."
13. On the rhetoric of the Anthropocene, also see Rickards (2015). On "materialized figurations," see Haraway (1994).
14. Perhaps, in Jennifer Ladino's (2015) terms, I simply was not the "implied tourist" for this place: the visitor embodying the intended norms and understandings of the site's designers and managers. Or perhaps my confusion was the result of having adopted the position of an environmental "killjoy," in Sara Ahmed's (2010, 39) sense of the term, "refusing to share an orientation towards certain things as being good," instead focusing on the negative dimensions that are widely "hidden, displaced, or negated." Thank you to Jennifer Hamilton for pointing out that I might be an "environmental killjoy" in a productive roundtable discussion between myself, Sara Ahmed, Sarah Franklin, Elspeth Probyn, and Eben Kirksey: "Happiness, Ecology, and Life in Glass," Environmental Humanities, University of New South Wales, March 24, 2015.
15. Having now spent much more time hanging around under crow roosts, I understand that this behavior is perfectly normal evening excitement.
16. On the surface, Derrida's notion of an unlimited hospitality might seem to offer an alternative way forward, a way in which hospitality might undermine its own appropriative foundation (see note 3). Pushed to the extreme, hospitality might escape some of the key problems I have been pointing to as the distinction between guest and host is blurred and problematized. With this in mind, perhaps we needn't abdicate the position of host but must rather take it up more absolutely?

However, I do not think that this is the right way forward for our Anthropocenic Earth. As an approach, hospitality papers over the "we": who has assumed the position of host, and at whose expense? Furthermore, to the extent that hospitality becomes "unlimited" it is no longer hospitality; to the extent that it does not, that it cannot, it remains grounded within an appropriative claim. In short, my position is that hospitality is the wrong approach.

17. For two quite different critical discussions of this approach to conservation, see Collard, Dempsey, and Sundberg (2015b) and Wuerthner, Crist, and Butler (2014).
18. There are a broad range of problems with the general approach laid out in *An Ecomodernist Manifesto*. The profound faith in technology and the market to overcome the biophysical limits of the Earth is startling, as is the reduction of sustainability to a question of energy and resource efficiency in a way that ignores the many social justice, animal welfare, and indeed broader environmental issues (for example, nuclear waste) associated with each of these processes. For a fuller critical discussion of this approach, see the assembled replies in *Environmental Humanities* 7 (2015).
19. Strategies for cohabitation cannot be developed in the abstract. They require situated engagement with local specificity. As such, the basic suggestions offered here may well be inappropriate in other parts of the world, especially those parts with fewer resources and less flexibility in meeting their human needs. At the same time, however, we ought not to assume that resource scarcity always makes cohabitation more difficult and that the path to "progress" is necessarily one of exerting greater and greater control over all others.
20. On this topic, see "Experimenting" and chapter 1, as well as van Dooren (2016). This openness to experimentation does not mean that we ought to *encourage* animals and plants to move around the world (although the assisted relocation of species in the face of climate change is a live and important topic). But when they do move, either through the accidental/careless action of people or their own seeking, and when they work to explore and develop modes of life for new places, we ought to respect those efforts as best we can.
21. This is an argument that moves along related lines to Mike Hulme's (2014) critique of geoengineering as a "solution" to climate change. Similarly, I would suggest that we need to insist that managing populations like these crows *cannot* be allowed to be the central issue for conservation and wildlife management.

FUMIGATING

1. As Clark and Yusoff (2014) propose in their "combustion-centric" analysis, thinking through fire might also draw us "beyond the human, beyond life, and even beyond the Earth."
2. Thank you to the person who suggested the term "Anthro-crow-cene" to me during one of my visits to the Rachel Carson Center at LMU Munich. Sadly, there were so many productive conversations during that period that I don't now know whose suggestion it was.
3. A fascinating noncorvid example of this is the way in which some birds of prey in Australia have been widely reported by Indigenous peoples and others to spread

fire deliberately—and with it the good hunting provided by fleeing animals—by picking up burning sticks and dropping them in an unburned area (Wilson 2016).

4. RECOGNIZING RAVENS: BECOMING SUBJECT TO EACH OTHER

1. Matthew Chrulew has written about the "unnamed naming" involved in (nonhuman) animal systems of identification, drawing on the work of Heini Hediger. Over the past several years, Chrulew has been engaged in a range of fascinating projects on Hediger's life and work. See, for example, Chrulew (2010, 2011b).
2. Some support for this possibility might be found in egg-aversion tools, which involve the use of tainted eggs to encourage members of one species not to eat the eggs of another species of, usually endangered, birds. Some success with such an approach has, for example, been achieved with another corvid predator, the Steller's jay (*Cyanocitta stelleri*), in efforts to conserve the marbled murrelet (*Brachyramphus marmoratus*) (Gabriel and Golightly 2011).
3. This account is based on an interview by the author with Roy Averill-Murray, Desert Tortoise Recovery Coordinator for the USFWS. The interview was conducted in Reno, Nevada, on October 26, 2016. For a written account of the key causes of the decline of the species, see Averill-Murray et al. (2012).
4. While the terminology of "lethal control" or "culling" is generally preferred by those involved in conservation, I have used the term "killing" throughout this chapter and this book. This is a deliberate choice. Killing is what is happening here, and like David Clark and Deborah Bird Rose, I see these other terms as alibis that allow us to sidestep the real significance of these actions, positioning this killing as inevitable, justified, something other than what it is (Clark 2004, 62; Rose 2011b, 27). I fully accept that in some cases killing wildlife is the best available option, but even in these cases it is still killing and must be struggled and lived with as such (van Dooren 2015).
5. Interviews with William (Bill) Boarman conducted by the author from October 31 to November 2, 2016, in Joshua Tree and Palm Springs, California.
6. Wildlife Services is a division of the U.S. Department of Agriculture (USDA).
7. Why this kind of proven "guilt" should make a difference is itself a fascinating question, but it is one that I will not be able to pursue here.
8. On the violences of a "balanced nature," see chapter 1.
9. On neophobia, see "Experimenting."
10. While these proposals raise a range of tantalizing possibilities for conservation, they are not without their dangers. It is, for example, hard not to make a connection here to the growing body of research on drone and other forms of remote warfare and surveillance. Despite the fact that the proposed interventions in the Mojave are nonlethal, many similar problems associated with the limitations of accurate identification, of distance and detachment, will likely be present, albeit in mutated forms (Gregory 2011, Suchman 2015, Kaplan 2017, Shaw 2017). To date, remote technologies for more standard forms of conservation monitoring and enforcement have been subject to very little ethical scrutiny, but here too there are contentious issues emerging (Lew 2015, Humle et al. 2014, Sandbrook 2015).

11. Likewise, these projects connect up with the military in interesting ways. Hardshell Labs is already collaborating with weapons designers on some of their projects, drawing on a talented pool of engineers who are happy to put their skills to work in a way that is both interesting and viewed as beneficial and largely benign. As Tim put it, the collaboration is "such a relief for them, psychically, just to know 'oh, we're just trying to save some turtles.'" The military is also an important potential source of funding for R&D projects, as it is interested in wildlife-management applications on the extensive lands that it manages, but it also seeks technologies with potential for adaptation to other military uses (a possibility driven home for me sitting next to an Air Force representative at the previously mentioned Raven Workshop).
12. As fascinating as AI and videogame proposals are, opening up their own questions of technologically mediated forms of recognition, they will not be explored in this chapter.
13. These three broad meanings are drawn from McQueen (n.d.).
14. Presentation by Tim Shields at USFWS Raven Workshop, Palm Springs, November 1–2, 2016.
15. These minimal notions of agency have played an important role in moving away from the dematerializing effects of the "linguistic turn" and long histories of viewing matter through lenses of "automatism or mechanism" (Bennet 2010, 3; Iovino and Oppermann 2012). Through this work we see that "objects" are not as lifeless or self-contained as has often been assumed and that the border(s) between subject and object are fuzzy and unstable. But such an approach also runs the risk of missing important differences, making it difficult to pay attention to the multiplicity of forms of acting, being, and becoming in/with the world. As Eduardo Kohn (2007, 5) notes, we require approaches to agency that "recognize that some nonhumans are selves. As such, they are not just represented but they also represent." In addition, it is important to note that the recognition of subjectivity has been hard won by many humans and nonhumans and continues to play vital (even if not always coherent or ideal) roles in the apportioning of ethical and political consideration.
16. I use the notion of a "self" to refer to a broader space of being composed of those who "respond" (van Dooren and Rose 2016). Such response, which can be found in the agency of plants and even cells, does not require subjectivity in the sense described here.
17. All quotations from Morizot's work are drawn from Despret (2016a), utilizing her translations.
18. On a related topic, see the discussion in chapter 1 of attending to others' experimental interjections as a ethicopolitical work of community. I understand the kind of diplomacy described in this chapter to involve a more cognitively demanding back and forth of proposals than is necessarily present in the more general space of experimentation described in chapter 1.
19. This work is grounded in a critical engagement with Hegel's foundational thought on recognition and its reinterpretation in the work of Jessica Benjamin.
20. Grounded in a broadly Foucauldian account of relations of power/knowledge, Judith Butler has explored the social, political, and cultural contexts within which recognition takes place, paying particular attention to the "norms of recognition" that shape the specific ways in which others are rendered legible and visible as

meaningful subjects (Butler 2004, 43; 1997). This is important work that might also be applied to the ways in which nonhuman animals have been rendered legible through scientific, philosophical, and broader cultural lenses, that is, how animal bodies emerge in and through processes of what Lynda Birke and colleagues (2004) call "animaling," to remind us that this category is an ongoing project with shifting meanings.

21. Understanding ravens as wily subjects and agents—as a species or set of populations in the desert, but also as individual birds—also requires us to rethink another common framing of the current situation, namely, as a "human problem." During my travels I frequently heard it said that the "real cause" of raven predation of desert tortoises was human activity. This framing is particularly popular with people opposing the killing of ravens and simplistic framings of the "raven problem," instead demanding that we focus our attention on the anthropogenic resources enabling their population growth. The danger with the "human problem" framing, however, is its capacity to reduce raven populations and behaviors to a simple question of resource abundance, as though there is only one way that ravens might be and that they reliably play out this script constrained only by external environmental conditions. While I applaud the effort not to cover over various forms of human complicity and responsibility, this should not become the whole story either. Current Mojave dynamics are the product of a complex and shifting "apparatus" (Despret 2008b) of human/raven/tortoise and more. They are not fixed once and for all: each party is becoming differently as a result of environmental changes, interactions with others, and the unique capacities and inheritances they bring with them. Holding onto, paying attention to, this complexity is the first step toward a diplomatic response.
22. In a related vein, drawing on Jamie Lorimer's recent work, we might think about these Hardshell Labs proposals as a strange new form of "probiotic" environmentality, an alternative to more conventional "antibiotic" efforts simply to dominate or eradicate inconvenient others. While there is often much to be celebrated in such a shift, Lorimer (2017, 3) also reminds us that these new approaches can easily sustain, rather than transform, "the unequal, proprietorial and anthropocentric character of prevalent forms of late modern biopolitics."
23. For a discussion of corvid caching and its connection in experimental research to understandings of a rich temporal awareness (in particular future-oriented, or prospective, cognition) see chapter 5. Also see "Stealing."
24. Biological discussions now often center on "theory-of-mind-like abilities" rather than the absence or presence of a ToM. This approach is grounded in the understanding that what we call ToM is perhaps a complex set of abilities and capacities present in various ways and to various degrees in different species, as well as in humans at different developmental stages. This more ambiguous label allows for more nuanced conversations about specific kinds of abilities/capacities associated with the recognition of others as subjects (Bugnyar 2007).
25. In Hegel's foundational work, recognition is at the heart of an explicit division between the human and the merely animal. While all animals, including humans, desire continuity of existence—*conatus* in the Spinozan sense—in humans this desire takes a particular form. As Hasana Sharp (2011, 128–29) puts it, in the case of humans, "Hegel contends that the desire to live is implicitly the desire for

recognition . . . what one experiences 'immediately' as the urge to persist becomes through the mediation of self-consciousness the desire for reflection in another self-consciousness." Also see Ikäheimo (2014). Even in Judith Butler's work, which challenges and redoes Hegel's thought in important ways, "the politics of recognition is necessarily humanist" (Sharp 2011, 151). For Butler, recognition is fundamentally a question of what it means to be properly *human*. In this way, Butler frequently collapses the notion of the subject with that of the "fully human." While the human is never given at the outset or settled once and for all, it remains the assumed limit of what will count as a valuable life, a grievable life, a life lived within the space of meaningful ethical relationship. In laying out the terrain in this way, Butler (1997, 2004) largely fails to direct critical attention toward the entrenched, naturalized norms of recognition that have long positioned animals and other nonhumans in precisely this devalued way.

26. I am not convinced that something like a theory of mind is *required* for a genuinely intersubjective encounter, although it perhaps re-forms what is possible and at stake here in some important ways.
27. In the work of Axel Honneth and others, grounded in what McQueen (2014) has called a "deficit model of recognition," the focus has been on contexts in which individuals lack recognition, or are misrecognized, in ways that produce injustices and various forms of harm. This approach, as McQueen argues, risks various kinds of universalizing and perhaps even essentializing about what it is to be human, and the basic human requirements of recognition, in a way that at least implies that there are value-neutral "correct" forms of recognition that we just need to get in place to ensure everyone's full "self-realisation." In contrast, Judith Butler and others have emphasized the ineradicable place of power relations, of contingent regimes of legibility and norms of recognition, in shaping which kinds of identities, which kinds of bodies, are made comfortable and indeed possible within given worlds.
28. Hegel emphasized the necessarily *mutual* nature of this recognition. In order for the other's modes of recognition to matter to us, to be formative, we must in turn recognize them as the kind of beings who matter. One of Hegel's principal concerns was with defining the social and political conditions necessary for each individual to achieve adequate recognition (Ikäheimo 2014, McQueen 2014).

GIFTING

1. As with so many of Lorenz's birds, this jackdaw was imprinted on him. On the ethics of imprinting with specific reference to Lorenz, see van Dooren (2014), chap. 4.
2. On interspecies friendships, see Lestel (2014a). On the specific possibilities for friendships between birds and their human feeders, see Kirksey et al. (2018).

5. PROVISIONING CROWS: CULTIVATING ECOLOGIES OF HOPE

1. Caching is a common behavior among the larger family Corvidae; see "Stealing."
2. *Sihek* is the Chamorro name for the Micronesian kingfisher (*Halcyon cinnamomina cinnamomina*), now extinct in the wild.

3. Interview with Fred Amidon, conducted in Honolulu on December 14, 2011. Amidon is an employee of the USFWS, Pacific Islands, and has been closely involved in the conservation of the aga.
4. See the afterword for a short discussion of this program.
5. For critical discussions of biogenetic approaches to Indigeneity in other contexts, see Kauanui (2008), Reardon (2005), and TallBear (2013).
6. Interview with Sarah Faegre.
7. Interviews with Beth Chagnon (Rota Division of Fish and Wildlife) and Lainie Zarones (CNMI Department of Lands and Natural Resources). Interviews conducted by the author in person and via Skype, respectively, in May and June 2016.
8. When a species meets certain requirements, it must be listed as endangered, triggering a range of other monitoring and compliance requirements. Similarly, where that species relies on a specific habitat area, it must be designated as "critical habitat." In the 2004 *Federal Register* entry on the declaration of critical habitat for the aga and two other local species, the USFWS notes that defending a steady stream of lawsuits—both from environmental groups claiming they're not doing enough and from corporations and other organizations effectively claiming that they're doing too much—now occupies most of their time and resources and undermines their capacity to engage in "adequate public participation" (USFWS, 2004). Putting the matter bluntly, they state: "The Service's present system for designating critical habitat has evolved since its original statutory prescription into a process that provides little real conservation benefit, is driven by litigation and the courts rather than biology, limits our ability to fully evaluate the science involved, consumes enormous agency resources, and imposes huge social and economic costs" (USFWS 2004, 62944). Yet year after year they remain bound within such a system, required by law to list a species as endangered and designate critical habitat for it, wherever the criteria are met.
9. Many others have pointed to one or more of these liminal aspects of hope. See, for example, Seeskin (2009).
10. This is not to say that many of these recruits are not also motivated by ideals like patriotism and the defense of freedom, as demonstrated powerfully in the 2017 documentary *Island Soldier* (Fitch 2017).
11. On the history of technologies for tracking wildlife, see Benson (2010). For a fascinating discussion of the ethics of bird monitoring, see Reinert (2013).
12. On the evolution of this general caching behavior, see "Stealing."
13. Nicola Clayton is a professor of comparative cognition at the University of Cambridge. She is one of the world's foremost authorities on prospective cognition. I interviewed her via telephone on July 12, 2016, and in person in Cambridge on June 9, 2017.
14. A recent study by Kabadayi and Osvath, focused on common ravens (*C. corax*), explored these birds' capacity to apply prospective cognition to "non-natural" behaviors, specifically requiring birds to save a token for future use in exchange for food. This study suggests that far from being restricted to particular behaviors, this capacity might be able to be applied generally (Kabadayi and Osvath 2017).
15. I am using the term "hope" differently than Donna Haraway (2016, 4) when she notes that "Alone, in our separate kinds of expertise and experience, we know both too much and too little, and so we succumb to despair or to hope, and neither is a

sensible attitude. Neither despair nor hope is tuned to the senses, to mindful matter, to material semiotics, to mortal earthlings in thick copresence." What Haraway here calls hope I have labeled (delusional) optimism. It is interesting also to note that Haraway here draws attention to these undesirable orientations as ones grounded in our being "alone." It is precisely this kind of solitude that the work of hope, as Jeff and others articulated it to me, challenges through its efforts to cultivate worlds with others.

16. A recent study published by Renee Ha and colleagues reached a similar conclusion. Drawing on both local community survey data and their long-term monitoring of the aga on Rota, they recommended changes to existing regulations so that only activities affecting lands that are *primary* crow nesting habitat require permitting. At the same time, they argued that this process must be made more transparent and straightforward and that some sort of monetary compensation should be established for those properties still affected by aga conservation (Sussman et al. 2015). This more streamlined approach will, however, become increasingly difficult as more and more species are listed as endangered on Rota.

REFERENCES

Adamson, Joni, William A. Gleason, and David N. Pellow, eds. 2016. *Keywords for Environmental Studies*. New York: New York University Press.

Adler, J. 2013. "Why Fire Makes Us Human." *Smithsonian Magazine* (June).

Aguon, J. 2011. "On Loving the Maps Our Hands Cannot Hold: Self-Determination of Colonized and Indigenous Peoples in International Law." *UCLA Asian Pacific American Law Journal* 16:47–73.

Ahmed, S. 2010. *The Promise of Happiness*. Durham, NC: Duke University Press.

Alaimo, S. 2016. *Exposed: Environmental Politics and Pleasures in Posthuman Times*. Minneapolis: University of Minnesota Press.

Anderson, B. 2006. "Transcending Without Transcendence: Utopianism and an Ethos of Hope." *Antipode* 38: 691–10.

Arista, N. 2010. "Navigating Uncharted Oceans of Meaning: Kaona as Historical and Interpretive Method." *PMLA* 125 (3): 663–69.

Asakawa-Haas, K., M. Schiestl, T. Bugnyar, and J. J. M. Massen. 2016. "Partner Choice in Raven (*Corvus corax*) Cooperation." *PLoS ONE* 11:1–15.

Åsberg, C. 2013. "The Timely Ethics of Posthumanist Gender Studies." *Feministische Studien* 31:7–12.

Averill-Murray, R. C., C. R. Darst, K. J. Field, and L .J. Allison. 2012. "A New Approach to Conservation of the Mojave Desert Tortoise." *BioScience* 62:893–99.

Bacchilega, C. 2007. *Legendary Hawaiʻi and the Politics of Place: Tradition, Translation, and Tourism*. State College: University of Pennsylvania Press.

Ball, D. L., I. Joaquin, L. Kaʻahaʻaina, M. Laut, and J. Nelson. 2016. "Restoring ʻAlalā to the Hawaiian Forest." U.S. Fish and Wildlife Service. https://www.fws.gov/endangered/news/episodes/bu-spring2016/story2/index.html.

Banko, P. C., D. L. Ball, and W. E. Banko. 2002. "Hawaiian Crow (*Corvus hawaiiensis*)." In *The Birds of North America Online*, ed. A. Poole. Ithaca, NY: Cornell Lab of Ornithology. https://birdsna.org/Species-Account/bna/species/hawcro/.

Banner, S. 2007. "Hawaii: Preparing to Be Colonized." In *Possessing the Pacific: Land, Settlers, and Indigenous People from Australia to Alaska*. Cambridge, MA: Harvard University Press.

Barad, K. 2007. *Meeting the Universe Halfway: Quantum Physics and the Entanglement of Matter and Meaning.* Durham, NC: Duke University Press.

——. 2010. "Quantum Entanglements and Hauntological Relations of Inheritance: Dis/continuities, SpaceTime Enfoldings, and Justice-to-Come." *Derrida Today* 3:240–68.

Barlow, C. 2000. *The Ghosts of Evolution: Nonsensical Fruit, Missing Partners, and Other Ecological Anachronisms*. New York: Basic.

Barua, M. 2014. "Volatile Ecologies: Towards a Material Politics of Human-Animal Relations." *Environment and Planning A* 46:1462–78.

Bastian, M. 2013. "Political Apologies and the Question of a 'Shared Time' in the Australian Context." *Theory, Culture, & Society*, preprint.

Bautista, K. 2018. "Power on Rota Almost Fully Restored." *Saipan Tribune*. September 18.

BBC. 2007. "Wild Crows Inhabiting the City Use It to Their Advantage." *BBC Wildlife*. https://www.youtube.com/watch?v=BGPGknpq3eo.

BCC (Brisbane City Council). 2014. "Torresian Crow." https://www.brisbane.qld.gov.au/environment-waste/natural-environment/protecting-wildlife-brisbane/living-wildlife/torresian-crow.

Bennet, J. 2010. *Vibrant Matter: A Political Ecology of Things*. Durham, NC: Duke University Press.

Bennett, J., and W. Connolly. 2012. "The Crumpled Handkerchief." In *Time and History in Deleuze and Serres*, ed. B. Herzogenrath, 153–73. London: Bloomsbury.

Benson, E. 2010. *Wired Wilderness: Technologies of Tracking and the Making of Modern Wildlife*. Baltimore, MD: Johns Hopkins University Press.

——. 2011. "Animal Writes: Historiography, Disciplinarity, and the Animal Trace." In *Making Animal Meaning*, ed. L. Kalof and G. M. Montgomery, 3–16). East Lansing: Michigan State University Press.

Berger-Tal, O., T. Polak, A. Oron, B. P. Kotler, and D. Saltz. 2011. "Integrating Animal Behavior and Conservation Biology: A Conceptual Framework." *Behavioral Ecology* 22.

Berlinger, J. 2017. "Caught in the Middle: The 4,000-Year-Old People Sharing an Island with the U.S. Military." *CNN Online*.

Birch, T. 2018. "'On What Terms Can We Speak?' Refusal, Resurgence and Climate Justice." *Coolabah* 24/25:2–16.

Birch, T. H. 1993. "Moral Considerability and Universal Consideration." *Environmental Ethics* 15 (4): 313–32.

BirdLife International. 2017. "EU Agriculture." http://www.birdlife.org/europe-and-central-asia/policy/eu-agriculture.

Birke, L., M. Bryld, and N. Lykke. 2004. "Animal Performances: An Exploration of Intersections Between Feminist Science Studies and Studies of Human/Animal Relationships." *Feminist Theory* 5:167–83.

Blaser, M. 2016. "Is Another Cosmopolitics Possible?" *Cultural Anthropology* 31:545–70.

Blaser, M., and M. de la Cadena. 2018. "Pluriverse: Proposals for a World of Many Worlds." In *A World of Many Worlds*, ed. M. Blaser and M. de la Cadena. Durham, NC: Duke University Press.

Boarman, W. I. 1993. "When a Native Predator Becomes a Pest: A Case Study." In *Conservation and Resource Management*, ed. S. K. Majumdar, 186–201. Philadelphia: Pennsylvania Academy of Science.

Boarman, W. I., and B. Heinrich. 1999. "Common Raven." In *The Birds of North America Online*, ed. A. Poole. Ithaca, NY: Cornell Lab of Ornithology.

Botkin, D. B. 1990. *Discordant Harmonies: A New Ecology for the Twenty-first Century.* Oxford: Oxford University Press.

Bowker, G. C. 2000. "Biodiversity Datadiversity." *Social Studies of Science* 30:643–83.

Boyer, A. G. 2008. "Extinction Patterns in the Avifauna of the Hawaiian Islands." *Diversity and Distributions* 14 (3): 509–17.

Bradshaw, C. J. A., and W. W. White. 2006. "Rapid Development of Cleaning Behaviour by Torresian Crows *Corvus orru* on Non-native Banteng *Bos javanicus* in Northern Australia." *Journal of Avian Biology* 37:409–11.

Bradshaw, C. J. A., Y. Isagi, S. Kaneko, D. J. M. S. Bowman, and B. W. Brook. 2006. "Conservation Value of Non-native Banteng in Northern Australia." *Conservation Biology* 20:1306–11.

Bradshaw, G. A., A. N. Schore, J. L. Brown, J. H. Poole, and C. J. Moss. 2005. "Social Trauma: Early Disruption of Attachment Can Affect the Physiology, Behaviour and Culture of Animals and Humans Over Generations." *Nature* 433:807.

Braidotti, R. 2013. *The Posthuman*. Cambridge: Polity.

Brockman, J. 2009. "We Are as Gods and Have to Get Good at It: Stewart Brand Talks About His Ecopragmatist Manifesto." *Edge*. https://www.edge.org/conversation/stewart_brand-we-are-as-gods-and-have-to-get-good-at-it.

Brooke, J. 2005a. "Apparel Factories in Saipan Are Threatened by End of Quotas." *New York Times*, April 12.

——. 2005b. "On Farthest U.S. Shores, Iraq Is a Way to a Dream." *New York Times*, July 31.

Brown, M. 2016. *Clever Crows: Investigating Behaviour and Learning in Wild Torresian Crows (*Corvus orru*) and Related Cracticids in a Suburban Environment.* Brisbane: Griffith University.

Brown, M. J. 2014. "Australian Crows." *The Corvid Blog*. http://coyot.es/thecorvidblog/tag/australian-crows/.

Brown, M. J., and D. N. Jones. 2016. "Cautious Crows: Neophobia in Torresian Crows (*Corvus orru*) Compared with Three Other Corvids in Suburban Australia." *Ethology* 122:726–33.

Brown, S. L. 2017. "Magpies Swooping ? You Should Try Making Friends with Them, Expert Says." *ABC News Online*. https://www.abc.net.au/news/2017-08-31/make-friends-with-magpies-to-avoid-swooping-expert-says/8856438.

Buchanan, B. 2008. *Onto-Ethologies: The Animal Environments of Uexküll, Heidegger, Merleau-Ponty, and Deleuze*. Albany, NY: SUNY Press.

Buchanan, B., M. Bastian, and M. Chrulew, eds. 2018. "Field Philosophy and Other Experiments." Special issue. *parallax* 24 (4).

Bugnyar, T. 2007. "An Integrative Approach to the Study of 'Theory-of-Mind'-Like Abilities in Ravens." *Japanese Journal of Animal Psychology* 57:15–27.

——. 2010. "Knower-Guesser Differentiation in Ravens: Others' Viewpoints Matter." *Proceedings of the Royal Society B, Published*: 1–7.

Bugnyar, T., and B. Heinrich. 2005. "Ravens, *Corvus corax*, Differentiate Between Knowledgeable and Ignorant Competitors." *Proceedings. Biological Sciences/The Royal Society* 272:1641–46.
——. 2006. "Pilfering Ravens: *Corvus corax* Adjust Their Behaviour to Social Context and Identity of Competitors." *Animal Cognition* 9:369–76.
Bugnyar, T., and K. Kotrschal. 2002. "Observational Learning and the Raiding of Food Caches in Ravens, *Corvus corax*: Is It 'Tactical' Deception?" *Animal Behaviour* 64:185–95.
Bugnyar, T., S. A. Reber, and C. Buckner. 2016. "Ravens Attribute Visual Access to Unseen Competitors." *Nature Communications* 7:1–6.
Bugnyar, T., M. Stöwe, and B. Heinrich. 2007. "The Ontogeny of Caching in Ravens, *Corvus corax*." *Animal Behaviour* 74:757–67.
Butler, J. 1993. *Bodies That Matter: On the Discursive Limits of "Sex."* New York: Routledge.
——. 1997. *The Psychic Life of Power: Theories in Subjection*. Stanford, CA: Stanford University Press.
——. 2004. *Precarious Life: The Powers of Mourning and Violence*. London: Verso.
Camacho, K. L. 2011. *Cultures of Commemoration: The Politics of War, Memory, and History in the Mariana Islands*. Honolulu: University of Hawai'i Press.
Candea, M. 2010. "'I Fell in Love with Carlos the Meerkat': Engagement and Detachment in Human-Animal Relations." *American Ethnologist* 37:241–58.
Candea, M. 2012. "Derrida en Corse? Hospitality as Scale-Free Abstraction." *Journal of the Royal Anthropological Institute* 18:S34–S48.
Cave, J. 2015. "The Pentagon Wants to Bomb the Hell Out of This Tiny Pacific Island." *Huffington Post*. https://www.huffingtonpost.com/2015/05/29/pagan-island-marines-military_n_7342168.html.
CEDSPC. 2009. *Comprehensive Economic Development Strategic Plan 2009–2014*. Saipan, CNMI: Commonwealth Economic Development Strategic Planning Commission, CNMI Department of Commerce.
Cheney, J., and A. Weston. 1999. "Environmental Ethics as Environmental Etiquette." *Environmental Ethics* 21:115–34.
Chrulew, M. 2010. "From Zoo to Zoopolis: Effectively Enacting Eden." *Metamorphoses of the Zoo: Animal Encounter After Noah*, ed. R. R. Acampora. Plymouth, UK: Lexington.
——. 2011a. "Managing Love and Death at the Zoo: The Biopolitics of Endangered Species Preservation." *Australian Humanities Review* 50:137–57.
——. 2011b. "Reflections in Philosophical Ethology." Paper presented at *The History, Philosophy and Future of Ethology*, Macquarie University, Australia.
——. 2017a. "Animals as Biopolitical Subjects." In *Foucault and Animals*, ed. M. Chrulew, D. J. Wadiwel, and L. Lawlor. Leiden: Brill.
——. 2017b. "Saving the Golden Lion Tamarin." In *Extinction Studies: Stories of Time, Death, and Generations*, ed. D. B. Rose, T. van Dooren, and M. Chrulew. New York: Columbia University Press.
CIA. 2016. "Northern Mariana Islands." *CIA World Factbook*. https://www.cia.gov/library/publications/the-world-factbook/geos/cq.html.
Clark, D. L. 2004. "On Being 'the Last Kantian in Nazi Germany': Dwelling with Animals After Levinas." In *Postmodernism and the Ethical Subject*, ed. B. Gabriel and S. Ilcan, 41–75. Montreal: McGill-Queen's University Press.

Clark, N. 2011. *Inhuman Nature: Sociable Life on a Dynamic Planet*. London: Sage.
——. 2014. "Geopolitics and the Disaster of the Anthropocene." *Sociological Review* 62:19–37.
Clark, N., and K. Yusoff. 2014. "Combustion and Society: A Fire-Centered History of Energy Use." *Theory, Culture & Society* 31:203–26.
Clarke, P. A. 2016. "Birds as Totemic Beings and Creators in the Lower Murray, South Australia." *Journal of Ethnobiology* 36 (2): 277–93.
Clayton, N. S. 2015a. "Feathered Apes Who Say Thanks with Shiny Trinkets." *New Scientist*.
——. 2015b. "Ways of Thinking: From Crows to Children and Back Again." *Quarterly Journal of Experimental Psychology* 68:209–41.
Clayton, N. S., T. J. Bussey, and A. Dickinson. 2003. "Can Animals Recall the Past and Plan for the Future?" *Nature Reviews Neuroscience* 4:685–91.
Clements, F. E. 1916. *Plant Succession*. Carnegie Institution of Washington publication 242.
Clifford, J. 1986. "Introduction: Partial Truths." In *Writing Culture: The Poetics and Politics of Ethnography*, ed. J. Clifford and G. E. Marcus. Berkeley: University of California Press.
Colebrook, C. 2012. "Not Symbiosis, Not Now: Why Anthropogenic Change Is Not Really Human." *Oxford Literary Review* 34:185–209.
Collard, R.-C., J. Dempsey, and J. Sundberg. 2015a. "A Manifesto for Abundant Futures." *Annals of the Association of American Geographers* 105:322–30.
——. 2015b. "The Moderns' Amnesia in Two Registers." *Environmental Humanities* 7:227–32.
Coulthard, G. S. 2014. *Red Skin, White Masks: Rejecting the Colonial Politics of Recognition*. Minneapolis: University of Minnesota Press.
Connolly, W. E. 2005. *Pluralism*. Durham, NC: Duke University Press.
Cooper, M. 2006. "Pre-empting Emergence: The Biological Turn in the War on Terror." *Theory, Culture & Society* 23:113–35.
Correia, S. P. C., A. Dickinson, and N. S. Clayton. 2007. "Western Scrub-Jays Anticipate Future Needs Independently of Their Current Motivational State." *Current Biology* 17:856–61.
Cosier, S. 2010. "Something to Crow About in Tokyo." *Audubon News*.
Crapanzano, V. 2003. "Reflections on Hope as a Category of Social and Psychological Analysis." *Cultural Anthropology* 18:3–32.
Crist, E. 1999. *Images of Animals: Anthropomorphism and Animal Mind*. Philadelphia: Temple University Press.
——. 2013a. "Ecocide and the Extinction of Animal Minds." In *Ignoring Nature No More: The Case for Compassionate Conservation*, ed. M. Bekoff. Chicago: University of Chicago Press.
——. 2013b. "On the Poverty of Our Nomenclature." *Environmental Humanities* 3:129–47.
Crutzen, P. J., and E. F. Stoermer. 2000. "The Anthropocene." *IGBP Newsletter* 41:17–18.
Csurhes, S. 2016. *Indian House Crow (*Corvus splendens*): Invasive Animal Risk Assessment*. Brisbane.
Culliney, S. M. 2011. "Seed Dispersal by the Critically Endangered Alala (*Corvus hawaiiensis*) and Integrating Community Values Into Alala (*Corvus hawaiiensis*) Recovery." MS thesis, Colorado State University.

Culliney, S. M., L. Pejchar, R. Switzer, and V. Ruiz-Guitierrez. 2012. "Seed Dispersal by a Captive Corvid: The Role of the 'Alala (*Corvus hawaiiensis*) in Shaping Hawai'i's Plant Communities." *Ecological Applications* 22:1718–32.
Cuomo, C. J. 1998. *Feminism and Ecological Communities: An Ethic of Flourishing*. London: Routledge.
D'Angelo, C. 2018. "Super Typhoon Yutu Shows Not All U.S. Cyclones Get Equal Treatment." *Huffington Post*, October 27.
Dally, J. M., N. S. Clayton, and N. J. Emery. 2006. "The Behaviour and Evolution of Cache Protection and Pilferage." *Animal Behaviour* 72:13–23.
Darwin, C. (1871) 1981. *The Descent of Man and Selection in Relation to Sex*. Princeton, NJ: Princeton University Press.
Davis, M. A., M. K. Chew, R. J. Hobbs, et al. 2011. "Don't Judge Species on Their Origins." *Nature* 474:153–54.
de Waal, F. B. M. 2008. "Putting the Altruism Back Into Altruism: The Evolution of Empathy." *Annual Review of Psychology* 59:279–300.
——. 2009. *Primates and Philosophers: How Morality Evolved*. Princeton, NJ: Princeton University Press.
DEHP. 2012. "Living with Crows." QLD Department of Environment and Heritage Protection. https://environment.des.qld.gov.au/wildlife/livingwith/crows/living-with-crows.html.
Delisle, C. T. 2015. "A History of Chamorro Nurse-Midwives in Guam and a 'Placental Politics' for Indigenous Feminism." *Intersections: Gender and Sexuality in Asia and the Pacific* 37.
DeLoughrey, E. 2017. "The Oceanic Turn: Submarine Futures of the Anthropocene." In *Humanities for the Environment: Integrating Knowledge, Forging New Constellations of Practice*, ed. J. Adamson and M. Davis. Abingdon: Routledge.
Derrida, J. 1995. *The Gift of Death*. Chicago: University of Chicago Press.
——. 1999. *Adieu to Emmanuel Levinas*. Stanford, CA: Stanford University Press.
——. 2000a. "Hospitality." *Angelaki: Journal of Theoretical Humanities* 5:3–18.
——. 2000b. *Of Hospitality*. Trans. R. Bowlby. Stanford, CA: Stanford University Press.
——. 2005. "The Principle of Hospitality." *Parallax* 11:6–9.
Derrida, J., E. Roudinesco, and J. Fort. 2004. *For What Tomorrow . . . A Dialogue*. Stanford, CA: Stanford University Press.
Despret, V. 2004. "The Body We Care for: Figures of Anthropo-zoo-genesis." *Body and Society* 10:111–34.
——. 2008a. "Culture and Gender Do Not Dissolve Into How Scientists 'Read' Nature: Thelma Rowell's Heterodoxy." *Rebels, Mavericks, and Heretics in Biology*, ed. Oren Harman and Michael R. Dietrich, 338–55. New Haven, CT: Yale University Press.
——. 2008b. "The Becomings of Subjectivity in Animal Worlds." *Subjectivity* 23:123–39.
——. 2014. "Domesticating Practices: "Domesticating Practices: The Case of Arabian Babblers." In *Routledge Handbook of Human-Animal Studies*, ed. Garry Marvin and Susan McHugh, 23–38. London: Routledge.
——. 2016a. *Experimenting Politics with Sheep: Ethology and the Art of Diplomacy*. Proceedings from Speculative Ethology 3. Perth.
——. 2016b. *What Would Animals Say If We Asked the Right Questions?* Minneapolis: University of Minnesota Press.

Devadas, V., and J. Mummery. 2007. "Community Without Community." *Borderlands* 6.
Diaz, V. 1994. "Simply Chamorro: Telling Tales of Demise and Survival in Guam." *The Contemporary Pacific* 6:29–58.
Dibley, B. 2012. "'The Shape of Things to Come': Seven Theses on the Anthropocene and Attachment." *Australian Humanities Review* 52:139–53.
Diprose, R. 2006. "Derrida and the Extraordinary Responsibility of Inheriting the Future-to-Come." *Social Semiotics* 16.
DLNR. 2012. *Ka'u Forest Reserve Management Plan*. Honolulu.
Driessen, C. 2014. "Animal Deliberation." In *Political Animals and Animal Politics*, ed. M. Wissenburg and D. Schlosberg. New York: Palgrave Macmillan.
DutchNews. 2013. "Supreme Court to Decide Fate of 23 Foreign Crows." *DutchNews.nl*, December 18.
EEA 2015. *State of Nature in the EU: Results from Reporting Under the Nature Directives 2007–2012*. EEA Technical Report 2/2015. Luxembourg.
Economist. 2011. "Welcome to the Anthropocene." *Economist*, May 26.
Ellis, E. 2011. "The Planet of No Return." *Breakthrough Journal* 2.
Emery, M. R., and A. R. Pierce. 2005. "Interrupting the Telos: Locating Subsistence in Contemporary US Forests." *Environment and Planning A* 37:981–93.
Emery, N. 2017. *Bird Brains: An Exploration of Avian Intelligence*. Princeton, NJ: Princeton University Press.
Emery, N., and N. Clayton. 2001. "Effects of Experience and Social Context on Prospective Caching Strategies by Scrub Jays." *Nature* 414:443–46.
Encinares, E. 2018. "Most Yutu Shelters Closed." *Saipan Tribune*, November 8.
Esposito, R. 2009. "Community and Nihilism." *Cosmos and History: The Journal of Natural and Social Philosophy* 5:24–36.
——. 2010. *Communitas: The Origin and Destiny of Community*. Stanford, CA: Stanford University Press.
Eugenio, H. V. 2018. "Humanitarian Crisis Looms After Super Typhoon Yutu Flattens Parts of Saipan and Tinian." *USA Today*, October 26.
Everding, S. E., and D. N. Jones. 2006. "Communal Roosting in a Suburban Population of Torresian Crows (*Corvus orru*)." *Landscape and Urban Planning* 74:21–33.
Fackler, M. 2008. "Japan Fights Crowds of Crows." *New York Times*, May 7.
Faegre, S. 2018. "Aga Release Blog Part 2: Super Typhoon Yutu." *San Diego Zoo Institute for Conservation Research Blog*, October 31.
Fairbrother, A., J. Purdy, T. Anderson, and R. Fell. 2014. "Risks of Neonicotinoid Insecticides to Honeybees." *Environmental Toxicology and Chemistry* 33:719–31.
Ferdowsian, H. R., D. L. Durham, C. Kimwele, et al. 2011. "Signs of Mood and Anxiety Disorders in Chimpanzees." *PLoS ONE* 6.
Fitch, N. 2017. "Island Soldier USA." Meerkat Media Collective. https://www.meerkatmedia.org/.
Fleming, S. 2010. "A Murder of Crows." Canadian Broadcast Corporation, Ontario.
Francis, R. A., and M. A. Chadwick. 2012. "What Makes a Species Synurbic?" *Applied Geography* 32:514–21.
Fraser, N. 2003. "Social Justice in the Age of Identity Politics: Redistribution, Recognition, and Participation." In *Redistribution or Recognition? A Political-Philosophical Exchange*, ed. F. Nancy and H. Axel. New York: Verso.

Fraser, O. N., and T. Bugnyar. 2010. "The Quality of Social Relationships in Ravens." *Animal Behaviour* 79:927–33.
——. 2012. "Reciprocity of Agonistic Support in Ravens." *Animal Behaviour* 83:171–77.
Friese, H. 2004. "Spaces of Hospitality." *Angelaki: Journal of the Theoretical Humanities* 9:67–79.
Fritts, T. H., and G. H. Rodda. 1998. "The Role of Introduced Species in the Degradation of Island Ecosystems: A Case History of Guam." *Annual Review of Ecology and Systematics* 29:113–40.
Fukami, T. 2015. "Historical Contingency in Community Assembly: Integrating Niches, Species Pools, and Priority Effects." *Annual Review of Ecology Evolution and Systematics* 46:1–23.
Gabriel, P. O., and R. T. Golightly. 2011. "Experimental Assessment of Taste Aversion Conditioning on Steller's Jays to Provide Potential Short-Term Improvement of Nest Survival of Marbled Murrelets in Northern California." Report to National Park Service. http://humboldt-dspace.calstate.edu/bitstream/handle/2148/877/Taste%20Aversion%20Conditioning%20Report%20FINAL%202011.pdf.
Garthwaite, J. 2013. "Mojave Mirrors: World's Largest Solar Plant Ready to Shine." *National Geographic*, July 27.
Geometrician Associates. 2012. *Final Environmental Assessment—Ka'u Forest Reserve Management Plan*. Honolulu, HI.
Gilbert, S. F., J. Sapp, and A. I. Tauber. 2012. "A Symbiotic View of Life: We Have Never Been Individuals." *Quarterly Review of Biology* 87:325–41.
Ginn, F. 2014. "Sticky Lives: Slugs, Detachment, and More-Than-Human Ethics in the Garden." *Transactions of the Institute of British Geographers* 39:532–44.
——. 2017. "Writing Anthroposcenes." Unpublished paper.
Ginn, F., U. Beisel, and M. Barua. 2014. "Flourishing with Awkward Creatures: Vulnerability, Togetherness, Killing." *Environmental Humanities* 4:113–23.
Giraud, E., G. Hollin, T. Potts, and I. Forsyth. 2018. "A Feminist Menagerie." *Feminist Review* 118 (1): 61–79.
Gleason, H. A. 1926. "The Individualistic Concept of the Plant Association." *Bulletin of the Torrey Botanical Club* 53:7.
Goldberg-Hiller, J., and N. K. Silva. 2011. "Sharks and Pigs: Animating Hawaiian Sovereignty Against the Anthropological Machine." *The South Atlantic Quarterly* 110:429–46.
Gon, S., III. "'O. Pua'a: Hawaiian Animal—or Forest Pest?" Unpublished paper.
Goodwin, D. 1986. *Crows of the World*. 2nd ed. London: British Museum.
Goodyear-Ka'ōpua, N. 2017. "Protectors of the Future, Not Protestors of the Past: Indigenous Pacific Activism and Mauna a Wākea." *South Atlantic Quarterly* 116 (1): 184–94.
Grebowicz, M., and H. Merrick. 2013. *Beyond the Cyborg: Adventures with Donna Haraway*. New York: Columbia University Press.
Green, K., and F. Ginn. 2014. "The Smell of Selfless Love: Sharing Vulnerability with Bees in Alternative Apiculture." *Environmental Humanities* 4:149–70.
Gregory, D. 2011. "From a View to a Kill: Drones and Late Modern War." *Theory, Culture & Society* 28:188–215.

Griffiths, P. E., and R. D. Gray. 2001. "Darwinism and Developmental Systems." In *Cycles of Contingency: Developmental Systems and Evolution*, ed. S. Oyama, P. E. Griffiths, and R. D. Gray. Cambridge, MA: MIT Press.

Grodzinski, U., and N. S. Clayton. 2010. "Problems Faced by Food-Caching Corvids and the Evolution of Cognitive Solutions." *Philosophical Transactions of the Royal Society of London Series B, Biological Sciences* 365:977–87.

Grosz, E. 2004. *The Nick of Time: Politics, Evolution, and the Untimely.* Durham, NC: Duke University Press.

Hadfield, M. G., and D. Haraway. In press. "The Tree-Snail Manifesto." *Current Anthropology.*

Hamilton, C. 2014. "The New Environmentalism Will Lead Us to Disaster." *Scientific American*, June 19.

Haraway, D. 1991. "Situated Knowledges: The Science Question in Feminism and the Privilege of Partial Perspective." In *Simians, Cyborgs, and Women: The Reinvention of Nature*. New York: Routledge.

——. 1992. "The Promises of Monsters: A Regenerative Politics for Inappropriate/d Others." In *Cultural Studies*, ed. L. Grossberg, C. Nelson, and P. A. Treichler. New York: Routledge.

——. 1994. "A Game of Cat's Cradle: Science Studies, Feminist Theory, Cultural Studies." *Configurations* 2 (1): 59–71.

——. 1997. *Modest_Witness@Second_Millennium.FemaleMan©_Meets_OncoMouse™: Feminism and Technoscience*. New York: Routledge.

——. 2008. *When Species Meet.* Minneapolis: University of Minnesota Press.

——. 2014. "SF: String Figures, Multispecies Muddles, Staying with the Trouble." Paper presented at University of Alberta, March 24.

——. 2015. "Anthropocene, Capitalocene, Plantationocene, Chthulucene: Making Kin." *Environmental Humanities* 6:159–65.

——. 2016. *Staying with the Trouble: Making Kin in the Chthulucene*. Durham, NC: Duke University Press.

Haraway, D., and T. N. Goodeve. 2000. *How Like a Leaf.* New York: Routledge.

Haring, E., B. Däubl, W. Pinsker, A. Kryukov, and A. Gamauf. 2012. "Genetic Divergences and Intraspecific Variation in Corvids of the Genus Corvus (Aves: Passeriformes: Corvidae)—a First Survey Based on Museum Specimens." *Journal of Zoological Systematics and Evolutionary Research* 50:230–46.

Harvey, G. 2006. *Animism: Respecting the Living World.* New York: Columbia University Press.

Hayward, I. 2006. "Why Do Crows Stand at Smoking Chimneys and Almost Sit in It and Preen Themselves?" Royal Society for the Protection of Birds. https://ww2.rspb.org.uk/birds-and-wildlife/bird-and-wildlife-guides/ask-an-expert/previous/chimneys.aspx.

Head, L. 2011. "Decentring 1788: Beyond Biotic Nativeness." *Geographical Research* 50:166–178.

Heinrich, B. 1988. "Why Do Ravens Fear Their Food?" *The Condor* 90:950–52.

——. 1999. *Mind of the Raven: Investigations and Adventures with Wolf-Birds.* New York: HarperCollins.

Heinrich, B., and T. Bugnyar. 2005. "Testing Problem Solving in Ravens: String-Pulling to Reach Food." *Ethology* 111:962–76.

——. 2007. "Just How Smart Are Ravens? *Scientific American* 296:64–71.
Heinrich, B., and J. Marzluff. 1995. "Why Ravens Share." *American Scientist* 83:342–49.
Hinchliffe, S., M. B. Kearnes, M. Degen, and S. Whatmore. 2005. "Urban Wild Things: A Cosmopolitical Experiment." *Environment and Planning D* 23:643–58.
Hinchliffe, S., and S. Whatmore. 2006. "Living Cities: Towards a Politics of Conviviality." *Science as Culture* 15:123–38.
ho'omanawanui, k. 2014. "Afterword: I ka 'Ōlelo ke Ola, In Words Is Life—Imagining the Future of Indigenous Literatures." In *The Oxford Companion of Literature of the Indigenous Americas*, ed. J. Cox and D. H. Justice, 675–82. Oxford: Oxford University Press.
Hofschneider, A. 2018. "The Super Typhoon American Media Forgot." *Columbia Journalism Review*, November 5.
Hollin, G., I. Forsyth, E. Giraud, and T. Potts. 2017. "Disentangling Barad: Materialisms and Ethics." *Social Studies of Science*. Forthcoming.
Honneth, A. 1995. *The Struggle for Recognition: The Grammar of Social Conflicts*. Cambridge: Polity.
Hulme, M. 2014. *Can Science Fix Climate Change: A Case Against Climate Engineering*. Cambridge: Polity.
Humle, T., R. Duffy, D. L. Roberts, F. A. V. S. John, and R. J. Smith. 2014. "Biology's Drones: Undermined by Fear." *Science* 344:1351.
Hurley, T. 2018. "More Alala to Be Set Free as Hawaiian Crow Soars." *Star Advertiser*, September 24.
Hutchison, K., and F. Jenkins, eds. 2013. *Women in Philosophy: What Needs to Change?* Oxford: Oxford University Press.
Ikäheimo, H. 2014. "Hegel's Concept of Recognition." In *Recognition: German Idealism as an Ongoing Challenge*, ed. C. Krijnen. Leiden: Brill.
Instone, L., and A. Taylor. 2015. "Thinking About Inheritance Through the Figure of the Anthropocene, from the Antipodes and in the Presence of Others." *Environmental Humanities* 7:133–50.
Iovino, S., and S. Oppermann. 2012. "Material Ecocriticism: Materiality, Agency, and Models of Narrativity." *Ecozon@* 3:75–91.
Iwashita, A. M. 2016. "Geothermal Potentials in Puna, Hawai'i: How Pele Teaches the Spaces Between." https://academiccommons.columbia.edu/doi/10.7916/D8CC17GS/download.
Jablonka, E., and M. J. Lamb. 2005. *Evolution in Four Dimensions: Genetic, Epigenetic, Behavioral, and Symbolic Variation in the History of Life*. Cambridge, MA: MIT Press.
James, I. 2010. "Naming the Nothing: Nancy and Blanchot on Community." *Culture, Theory, and Critique* 51:171–87.
Janzen, D. H., and P. S. Martin. 1982. "Neotropical Anachronisms: The Fruits the Gomphotheres Ate." *Science* 215:19–27.
Jones, D. 2018. *The Birds at My Table: Why We Feed Wild Birds and Why It Matters*. Ithaca, NY: Cornell University Press.
Kabadayi, C., and M. Osvath. 2017. "Ravens Parallel Great Apes in Flexible Planning for Tool-Use and Bartering." *Science* 357:202–4.

Kaplan, C. 2017. "Drone-o-Rama: Troubling the Temporal and Spatial Logics of Drone Warfare." In *Life in the Age of Drone Warfare*, ed. L. Parks and C. Kaplan. Durham, NC: Duke University Press.
Kareiva, P., R. Lalasz, and M. Marvier. 2011. "Conservation in the Anthropocene: Beyond Solitude and Fragility." *Breakthrough Journal* 2:29–37.
Kath, J., M. Maron, and P. K. Dunn. 2009. "Interspecific Competition and Small-Bird Diversity in an Urbanizing Landscape." *Landscape and Urban Planning* 92 (2): 72–79.
Kauanui, J. K. 2008. *Hawaiian Blood: Colonialism and the Politics of Sovereignty and Indigeneity*. Durham, NC: Duke University Press.
——. 2009. "Hawaiian Independence and International Law." *Indigenous Politics on WESU Radio*. Episode 23.
Kearnes, M., and T. van Dooren. 2017. "Rethinking the Final Frontier: Cosmo-Logics and an Ethic of Interstellar Flourishing." *GeoHumanities* 3:1–20.
Kelleher, J. S. (2018). "Typhoon Yutu Damage Sends Tourists Fleeing from Northern Mariana Islands." *Time* (October 30).
Kirch, P. V. 2011. "When Did the Polynesians Settle Hawaii? A Review of 150 Years of Scholarly Inquiry and a Tentative Answer." *Hawaiian Archaeology* 12:3–26.
Kirksey, E., P. Munro, T. van Dooren, et al. 2018. "Feeding the Flock: Wild Cockatoos and Their Facebook Friends." *Environment and Planning E: Nature and Space* 1 (4).
Kirksey, S. E. 2012. "Living with Parasites in Palo Verde National Park." *Environmental Humanities* 1:23–55.
——. 2015. *Emergent Ecologies*. Durham, NC: Duke University Press.
Kirksey, S. E., N. Shapiro, and M. Brodine. 2013. "Hope in Blasted Landscapes." *Social Science Information* 52:228–56.
Klump, B. C., J. E. M. V. D. Wal, J. J. H. S. Clair, and C. Rutz. 2015. "Context-Dependent 'Safekeeping' of Foraging Tools in New Caledonian Crows." *Proceedings of the Royal Society B: Biological Sciences* 282.
Knight, R. L., D. J. Grout, and S. A. Temple. 1987. "Nest-Defense Behavior of the American Crow in Urban and Rural Areas." *Condor* 89:175–77.
Kohn, E. 2007. "How Dogs Dream: Amazonian Natures and the Politics of Transspecies Engagement." *American Ethnologist* 34:1425–1548.
Kricher, J. 2009. *The Balance of Nature: Ecology's Enduring Myth*. Princeton, NJ: Princeton University Press.
Kuwada, B. K. 2015. "We Live in the Future. Come Join Us." https://hehiale.wordpress.com/2015/04/03/we-live-in-the-future-come-join-us/.
Ladino, J. K. 2015. "Mountains, Monuments, and Other Matter: Environmental Affects at Manzanar." *Environmental Humanities* 6:131–57.
Langford, J. 2017. "Avian Bedlam: Toward a Biosemiosis of Troubled Parrots." *Environmental Humanities* 9 (1): 84–107.
Latour, B. 1986. "The Powers of Association." In *Power, Action, and Belief: A New Sociology of Knowledge?*, ed. J. Law. London: Routledge and Kegan Paul.
——. 2004. *Politics of Nature: How to Bring the Sciences into Democracy*. Cambridge, MA: Harvard University Press.
Law, J. 2015. "What's Wrong with a One-World World." *Distinktion: Journal of Social Theory* 16 (1): 126–39.

Lawlor, L. 2007. *This is Not Sufficient: An Essay on Animality and Human Nature in Derrida*. New York: Columbia University Press.
LeMenager, S. 2017. *Climate Change and the Quest for Transformative Fictions*. Proceedings from Hacking the Anthropocene II: Weathering. University of Sydney.
Lenzen, M., D. Moran, K. Kanemoto, et al. 2012. "International Trade Drives Biodiversity Threats in Developing Nations." *Nature* 486:109–12.
Leonard, D. L. J. 2008. "Recovery Expenditures for Birds Listed Under the U.S. Endangered Species Act: The Disparity Between Mainland and Hawaiian Taxa." *Biological Conservation* 141:2054–61.
Leong, J., J. J. Marra, M. L. Finucane, et al. 2014. "Hawai'i and U.S. Affiliated Pacific Islands." In *Climate Change Impacts in the United States: The Third National Climate Assessment*, ed. J. M. Melillo, Terese (T. C.) Richmond, and G. W. Yohe, 537–56. Washington, DC: U.S. Global Change Research Program.
Lestel, D. 2014a. "The Friends of My Friends." *Angelaki* 19:133–47.
——. 2014b. "The Question of the Animal Subject." *Angelaki* 19:113–25.
Lestel, D., F. Brunois, and F. Gaunet. 2006. "Etho-Ethnology and Ethno-Ethology." *Social Science Information* 45:155–77.
Lew, J. 2015. "The Ethics of Drones in the Wild." *Mother Nature Network*. https://www.mnn.com/green-tech/gadgets-electronics/stories/the-ethics-of-drones-in-the-wild.
Lim, L. 2010. "As the Crow Flies, Tokyo Battles Avian Pest." *National Public Radio*. https://www.npr.org/templates/story/story.php?storyId=122291084.
Lorenz, K. 1952. *King Solomon's Ring: New Light on Animal Ways*. London: Routledge.
Lorimer, H. 2013. "Scaring Crows." *Geographical Review* 103:177–89.
Lorimer, J. 2016. "Gut Buddies." *Environmental Humanities* 8:57–76.
——. 2017. "Probiotic Environmentalities: Rewilding with Wolves and Worms." *Theory, Culture & Society* 34 (4).
Lorimer, J., and C. Driessen. 2014. "Wild Experiments at the Oostvaardersplassen: Rethinking Environmentalism in the Anthropocene." *Transactions of the Institute of British Geographers* 39:169–81.
Low, T. 2002. *The New Nature: Winners and Losers in Wild Australia*. Camberwell, Victoria: Penguin.
MacKenzie, E. 2016. "Online Map of Vancouver Crow Attacks: A New Tool Developed by Langara Instructors." *Vancouver Sun*. https://vancouversun.com/news/local-news/online-map-of-vancouver-crow-attacks-a-new-tool-developed-by-langara-instructors.
Maffi, L. 2004. "Maintaining and Restoring Biocultural Diversity: The Evolution of a Role for Ethnobiology." In *Ethnobotany and Conservation of Biocultural Diversity*, ed. T. J. S. Carlson and L. Maffi. Bronx, NY: New York Botanical Garden Press.
Maggs, D., and J. Robinson. 2016. "Recalibrating the Anthropocene: Sustainability in an Imaginary World." *Environmental Philosophy* 13:175–94.
Maly, K., and O. Maly. 2004. *He Mo'Olelo 'Āina: A Cultural Study of The Manukā Natural Area Reserve Lands of Manukā, District of Ka'ū and Kaulanamauna, District Of Kona, Island of Hawai'i*. Hilo, HI.
Maly, K., B. K. Pang, and C. P. M. Burrows. 2007. "Pigs in Hawai'i, from Traditional to Modern." Unpublished paper (on file with author).
Margalit, A. 2001. "Recognizing the Brother and the Other." *Proceedings of the Aristotelian Society, Supplementary Volume* 75:127–39.

Maron, M., M. J. Grey, C. P. Catterall, et al. 2013. "Avifaunal Disarray Due to a Single Despotic Species." *Diversity and Distributions* 19:1468–79.

Martin, G., D. Mincyte, and U. Münster. 2012. "Why Do We Value Diversity? Biocultural Diversity in a Global Context." *RCC Perspectives.*

Marzluff, J. M., K. J. McGowan, R. Donnelly, and R. L. Knight. 2001. "Causes and Consequences of Expanding American Crow Populations." In *Avian Ecology and Conservation in an Urbanizing World*, ed. J. Marzluff, R. Bowman, and R. Donnelly, 331–63. New York: Springer.

Marzluff, J. M., and T. Angell. 2005. *In the Company of Crows and Ravens.* New Haven, CT: Yale University Press.

——. 2012. *Gifts of the Crow: How Perception, Emotion, and Thought Allow Smart Birds to Behave Like Humans.* New York: Free Press.

Marzluff, J. M., and M. L. Miller. 2014. "Crows and Crow Feeders: Observations on Interspecific Semiotics." In *Biocommunication of Animals*, ed. Guenther Witzany, 233–47. Dordrecht: Springer Netherlands.

Marzluff, J. M., J. Walls, H. N. Cornell, et al. 2010. "Lasting Recognition of Threatening People by Wild American Crows." *Animal Behaviour* 79:699–707.

Massen, J. J. M., C. Ritter, and T. Bugnyar. 2015. "Tolerance and Reward Equity Predict Cooperation in Ravens (*Corvus corax*)." *Scientific Reports* 5:1–11.

Mauss, M. 1970. *The Gift: Forms and Functions of Exchange in Archaic Societies.* London: Cohen and West.

Mawyer, A., and J. K. Jacka. 2018. "Sovereignty, Conservation, and Island Ecological Futures." *Environmental Conservation* 45 (3): 238–51.

McCarthy, J. 2002. "First World Political Ecology: Lessons from the Wise Use Movement." *Environment and Planning A* 34:1281–1302.

McDougall, B. N. 2016. *Finding Meaning: Kaona and Contemporary Hawaiian Literature.* Tucson: University of Arizona Press.

McFall-Ngai, M., M. G. Hadfield, T. C. G. Bosch, et al. 2013. "Animals in a Bacterial World, a New Imperative for the Life Sciences." *Proceedings of the National Academy of Sciences of the United States of America* 110:3229–36.

McLauchlan, L. 2018. *Wild Cares: On Hedgehogs, Killing, and Kindness.* Sydney: University of New South Wales.

McNeill, J. R., and P. Engelke. 2016. *The Great Acceleration: An Environmental History of the Anthropocene Since 1945.* Cambridge, MA: Harvard University Press.

McQueen, P. N.d. "Social and Political Recognition." *Internet Encyclopedia of Philosophy.* https://www.iep.utm.edu/recog_sp/.

McQueen, P. 2014. "Honneth, Butler, and the Ambivalent Effects of Recognition." *Res Publica* 21:43–60.

Mol, A. 2002. *The Body Multiple: Ontology in Medical Practice.* Durham, NC: Duke University Press.

Moore, J. W. 2014. "The Capitalocene: Part 1: On the Nature and Origins of Our Ecological Crisis." Unpublished paper, Fernand Braudel Center, Binghamton University.

——. 2015. *Capitalism in the Web of Life: Ecology and the Accumulation of Capital.* New York: Verso.

Morin, M. E. 2006. "Putting Community Under Erasure: Derrida and Nancy on the Plurality of Singularities." *Culture Machine* 8:1–7.

Morizot, B. 2016. *Les diplomates: cohabiter avec les loups sur une nouvelle carte du vivant.* Paris: WildProjects Editions.
Mouffe, C. 1999. "Deliberative Democracy or Agonistic Pluralism? *Social Research* 66:745–58.
Muir, C. 2014. *The Broken Promise of Agricultural Progress: An Environmental History.* London: Routledge.
Munro, G. 1944. *Birds of Hawaii.* Honolulu: Tongg.
Münster, U. 2014. "Working for the Forest: The Ambivalent Intimacies of Human–Elephant Collaboration in South Indian Wildlife Conservation." *Ethnos: Journal of Anthropology* 81:425–47.
Murphy, M. 2015. "Unsettling Care: Troubling Transnational Itineraries of Care in Feminist Health Practices." *Social Studies of Science* 45:717–37.
Nākoa Oliveira, K.-A. R. K. 2014. *Ancestral Places: Understanding Kanaka Geographies.* Corvallis: Oregon State University Press.
Nancy, J.-L. 1991. *The Inoperative Community.* Minneapolis: University of Minnesota Press.
National Research Council. 1997. *The Scientific Bases for the Preservation of the Mariana Crow.* Washington, DC: National Academies Press.
Neimanis, A. 2016. *Bodies of Water: Posthuman Feminist Phenomenology.* London: Bloomsbury Academic.
Nihei, Y. 1995. "Variations of Behaviour of Carrion Crows *Corvus corone* Using Automobiles as Nutcrackers." *Japanese Journal of Ornithology* 44:21–35.
Nihei, Y., and H. Higuchi. 2001. "When and Where Did Crows Learn to Use Automobiles as Nutcrackers?" *Tohoku Psychologica Folia* 60:93–97.
NPS. 2005. *Reconnaissance Survey Significant Natural Areas and Cultural Sites: Island of Rota, Commonwealth of the Northern Mariana Islands.* Honolulu.
Nyari, A., C. Ryall, and A. T. Peterson. 2006. "Global Invasive Potential of the House Crow *Corvus splendens* Based on Ecological Niche Modelling." *Journal of Avian Biology* 37:306–11.
O'Gorman, E. 2012. *Flood Country: Floods in the Murray and Darling River Systems, 1850 to the Present.* Collingwood, Victoria: CSIRO.
——. 2014. "Remaking Wetlands: Rice Fields and Ducks in the Murrumbidgee River Region, NSW." In *Rethinking Invasion Ecologies from the Environmental Humanities*, ed. Jodi Frawley and Iain McCalman, 215–38. London: Routledge.
——. 2017. "Imagined Ecologies: The Spectre of Malaria in the Murrumbidgee Irrigation Area, 1919–45." *Environmental History* 22 (3): 486–514.
O'Gorman, E., and T. van Dooren. 2016. "The Promises of Pests: Wildlife in Agricultural Landscapes." *Australian Zoologist* 39 (1): 81–84.
OED. 1989. *The Oxford English Dictionary.* 2nd ed. Online resource.
Oliver, K. 2009. *Animal Lessons: How They Teach Us to Be Human.* New York: Columbia University Press.
Osorio, J. K. K. 2002. *Dismembering Lahui: A History of the Hawaiian Nation to 1887.* Honolulu: University of Hawai'i Press.
——, ed. 2014. *i ula i ka 'āina (land).* Honolulu: University of Hawai'i Press.
Osteen, M., ed. 2002. *The Question of the Gift: Essays Across Disciplines.* London: Routledge.

Ostojic, L., R. C. Shaw, L. G. Cheke, and N. S. Clayton. 2013. "Evidence Suggesting That Desire-State Attribution May Govern Food Sharing in Eurasian Jays." *Proceedings of the National Academy of Sciences of the United States of America* 110:4123–28.

Ottens, G., and C. Ryall. 2003. "House Crows in the Netherlands and Europe." *Dutch Birding* 25:312–19.

Oyama, S., P. E. Griffiths, and R. D. Gray. 2001. *Cycles of Contingency: Developmental Systems and Evolution*. Cambridge, MA: MIT Press.

Pacini-Ketchabaw, V., and F. Nxumalo. 2015. "Unruly Raccoons and Troubled Educators: Nature/Culture Divides in a Childcare Centre." *Environmental Humanities* 7:151–68.

Palmer, C. 2001. "Taming the Wild Profusion of Existing Things? A Study of Foucault, Power, and Human/Animal Relationships." *Environmental Ethics* 23:339–58.

——. 2003. "Colonization, Urbanization, and Animals." *Philosophy & Geography* 6:47–58.

——. 2010. *Animal Ethics in Context*. New York: Columbia University Press.

Papadopoulos, D. 2014. "Generation M. Matter, Makers, Microbiomes: Compost for Gaia." *Teknokultura* 11:637–45.

——. 2018. *Experimental Practice: Technoscience, Alterontologies, and More-Than-Social Movements*. Durham, NC: Duke University Press.

Patton, P. 2004. "Politics." In *Understanding Derrida*, ed. J. Roffe and J. Reynolds. London: Continuum.

Perez, C. S. 2014. "Ginen the Micronesian Kingfisher [*i sihek*]." In *From Unincorporated Territory [guma']*. Omnidawn.

Peterson, J. 2016. "American Soil, Chamorro Soul." Guam: The Guam Culture Guide, Inc.

Plumwood, V. 1993. *Feminism and the Mastery of Nature*. London: Routledge.

——. 2002. *Environmental Culture: The Ecological Crisis of Reason*. London: Routledge.

——. 2008. "Shadow Places and the Politics of Dwelling." *Ecological Humanities, Australian Humanities Review* 44.

——. 2009. "Nature in the Active Voice." *Australian Humanities Review* 46:113–29.

Pratt, M. L. 2003. *Imperial Eyes: Travel Writing and Transculturation*. Taylor and Francis.

Preston, S. D., and F. B. M. de Waal. 2005. "Empathy: Its Ultimate and Proximate Bases." *Behavioral and Brain Sciences* 25:1–72.

Puig de la Bellacasa, M. 2012. "'Nothing Comes Without Its World': Thinking with Care." *Sociological Review* 60:197–216.

——. 2017. *Matters of Care: Speculative Ethics in More Than Human Worlds*. Minneapolis: University of Minnesota Press.

Raby, C. R., D. M. Alexis, A. Dickinson, and N. S. Clayton. 2007. "Planning for the Future by Western Scrub-Jays." *Nature* 445:919–21.

Raby, C. R., and N. S. Clayton. 2009. "Prospective Cognition in Animals." *Behavioural Processes* 80:314–24.

Reardon, J. 2005. *Race to the Finish: Identity and Governance in an Age of Genomics*. Princeton, NJ: Princeton University Press.

Reinert, H. 2013. "The Care of Migrants: Telemetry and the Fragile Wild." *Environmental Humanities* 3:1–24.

——. 2016. "About a Stone: Some Notes on Geological Conviviality." *Environmental Humanities* 8.

Restall Orr, E. 2012. *The Wakeful World: Animism, Mind, and the Self in Nature.* Alresford, UK: John Hunt.

Rickards, L. A. 2015. "Metaphor and the Anthropocene: Presenting Humans as a Geological Force." *Geographical Research* 53.

Ristroph, E. B. 2007. "The Survival of Customary Law in the Northern Mariana Islands." *Chicago-Kent Journal of International and Comparative Law* 8:32–65.

Robbins, P. 2004. "Comparing Invasive Networks: Cultural and Political Biogeographies of Invasive Species." *Geographical Review* 94:139–56.

Robinette Ha, R., P. Bentzen, J. Marsh, and J. C. Ha. 2003. "Kinship and Association in Social Foraging Northwestern Crows (*Corvus caurinus*)." *Bird Behavior* 15:65–75.

Robinette, R. L., and J. C. Ha. 2001. "Social and Ecological Factors Influencing Vigilance by Northwestern Crows, *Corvus caurinus*." *Animal Behaviour* 62:447–52.

Robinette Ha, R., and J. C. Ha. 2003. "Effects of Ecology and Prey Characteristics on the Use of Alternative Social Foraging Tactics in Crows, *Corvus caurinus*." *Animal Behaviour* 66:309–16.

Roelvink, G., and M. Zolkos. 2015. "Affective Ontologies: Posthumanist Perspectives on the Self, Feeling, and Intersubjectivity." *Emotion, Space and Society* 14:47–49.

Rolston, H., III. 1999. "Respect for Life: Counting What Singer Finds of No Account." In *Singer and His Critics*, ed. D. Jamieson, 247–68. Oxford: Blackwell.

Rose, D. B. 2004. *Reports from a Wild Country: Ethics for Decolonisation.* Sydney: University of New South Wales Press.

——. 2006. "What If the Angel of History Were a Dog?" *Cultural Studies Review* 12:67–78.

——. 2007. "Recursive Epistemologies and an Ethics of Attention." In *Extraordinary Anthropology: Transformations in the Field*, ed. B. Miller, 88–102. Lincoln: University of Nebraska Press.

——. 2008. "Judas Work: Four Modes of Sorrow." *Environmental Philosophy* 5:51–66.

——. 2011a. "Flying Fox: Kin, Keystone, Kontaminant." *Australian Humanities Review* 50:119–36.

——. 2011b. *Wild Dog Dreaming: Love and Extinction.* Charlottesville: University of Virginia Press.

——. 2012a. "Cosmopolitics: The Kiss of Life." *New Formations* 76:101–13.

——. 2012b. "Multispecies Knots of Ethical Time." *Environmental Philosophy* 9:127–40.

——. 2013a. "In the Shadow of All This Death." In *Animal Death*, ed. J. Johnston and F. Probyn-Rapsey. Sydney: Sydney University Press.

——. 2013b. "Val Plumwood's Philosophical Animism: Attentive Inter-Actions in the Sentient World." *Environmental Humanities* 3:93–109.

Rose, D. B., and T. van Dooren. 2016. "Encountering a More-Than-Human World: Ethos and the Arts of Witness." In *Routledge Companion to the Environmental Humanities*, 120–28. London: Routledge.

Rosenberg, D., and S. Harding. 2005. "Introduction: Histories of the Future." In *Histories of the Future*, ed. D. Rosenberg and S. Harding. Durham, NC: Duke University Press.

Rosenthal, E. 2009. "Smuggling Europe's Waste to Poorer Countries." *New York Times*, September 26.
Roughgarden, J. 2009. "Is There a General Theory of Community Ecology?" *Biology & Philosophy* 24:521–29.
Routley, R., and V. Routley. 1979. "Against the Inevitability of Human Chauvinism." In *Ethics and Problems of the Twenty-First Century*, ed. Kenneth E. Goodpaster and Kenneth M. Sayer, 36–59. Notre Dame, IN: University of Notre Dame Press.
Rowell, T. E. 1974. "The Concept of Social Dominance." *Behavioral Biology* 11:131–54.
Rowley, I. 1973. "The Comparative Ecology of Australian Corvids, Parts 1–6." *CSIRO Wildlife Research* 18:1–169.
Ryall, C. 2003. "Notes on Ecology and Behaviour of House Crows at Hoek van Holland." *Dutch Birding* 25:167–71.
Ryall, J. 2008. "Bees Enlisted to Attack Crows in Tokyo." *National Geographic News*, July 14.
Sandbrook, C. 2015. "The Social Implications of Using Drones for Biodiversity Conservation." *Ambio* 44:636–47.
Sandilands, C. 2018. *Feminist Botany for the Age of Man*. Proceedings from the Sydney Environmental Humanities Lecture Series, Sydney.
Schilthuizen, M. 2018. *Darwin Comes to Town: How the Urban Jungle Drives Evolution*. New York: Picador.
Schloegl, C., K. Kotrschal, and T. Bugnyar. 2007. "Gaze Following in Common Ravens, *Corvus corax*: Ontogeny and Habituation." *Animal Behaviour* 74:769–78.
Schlosberg, D., and R. Coles. 2016. "The New Environmentalism of Everyday Life: Sustainability, Material Flows, and Movements." *Contemporary Political Theory* 15:160–81.
Schneider, G. W., and R. Winslow. 2014. "Parts and Wholes: The Human Microbiome, Ecological Ontology, and the Challenges of Community." *Perspectives in Biology and Medicine* 57:208–23.
Schwarz, J. 2003. "Crows Alter Their Thieving Behavior When Dealing with Kin, Other Birds." *UW Today*, March 11.
Seed, A. M., N. Clayton, and N. Emery. 2007. "Postconflict Third-Party Affiliation in Rooks, *Corvus frugilegus*." *Current Biology* 17:152–58.
——. 2008. "Cooperative Problem Solving in Rooks (*Corvus frugilegus*)." *Proceedings of the Royal Society B: Biological Sciences* 275:1421–29.
Seeskin, K. 2009. "The Promise and Problems of Hope." In *Hope in the Twenty-First Century*, ed. J. L. Hochheimer. Oxford: Inter-Disciplinary Press.
Serres, M. 2011. *Malfeasance: Appropriation Through Pollution?* Trans. A.-M. Feenberg-Dibon. Stanford, CA: Stanford University Press.
Serres, M., B. Latour, and R. Lapidus. 1995. *Conversations on Science, Culture, and Time*. Ann Arbor: University of Michigan Press.
Seuss, D. 1999. *The Lorax*. London: Harper Collins Children's Books.
Sewall, K. 2015. "The Girl Who Gets Gifts from Birds." *BBC News*. https://www.bbc.com/news/magazine-31604026.
Seymour, M., and J. Wolch. 2009. "Toward Zoöpolis? Innovation and Contradiction in a Conservation Community." *Journal of Urbanism: International Research on Placemaking and Urban Sustainability* 2:215–36.

Sharp, H. 2011. *Spinoza and the Politics of Renaturalization*. Chicago: University of Chicago Press.
Shaw, I. G. R. 2017. "The Great War of Enclosure: Securing the Skies." *Antipode* 49:883–906.
Shenon, P. 1993. "Made in the U.S.A.? Hard Labor on a Pacific Island: A Special Report: Saipan Sweatshops Are No American Dream." *New York Times*, July 18.
Shotwell, A. 2016. *Against Purity: Living Ethically in Compromised Times*. Minneapolis: University of Minnesota Press.
Silva, N. K. 2004. *Aloha Betrayed: Native Hawaiian Resistance to American Colonialism*. Durham, NC: Duke University Press.
Simberloff, D. 2004. "Community Ecology: Is It Time to Move on? (An American Society of Naturalists Presidential Address)." *American Naturalist* 163:787–99.
Singer, P. 1975. *Animal Liberation: A New Ethics for our Treatment of Animals*. New York: New York Review/Random House.
Slaterus, R., B. Aarts, and L. V. D. Bremer. 2009. *De Huiskraai in Nederland: risicoanalyse en beheer*. Beek-Ubbergen: SOVON, on behalf of Team Invasieve Exoten van het Ministerie van Landbouw, Natuur en Voedselkwaliteit.
Sodikoff, G. M. 2013. "The Time of Living Dead Species: Extinction Debt and Futurity in Madagascar." In *Debt: Ethics, the Environment, and the Economy*, ed. P. Y. Paik and M. Wiesner-Hanks. Bloomington: Indiana University Press.
Soulé, M. 1990. "The Onslaught of Alien Species, and Other Challenges in the Coming Decades." *Conservation Biology* 4:233–39.
Steadman, D. W. 2006. *Extinction and Biogeography of Tropical Pacific Birds*. Chicago: University of Chicago Press.
Steffen, W., P. J. Crutzen, and J. R. McNeill. 2007. "The Anthropocene: Are Humans Now Overwhelming the Great Forces of Nature." *Ambio: A Journal of the Human Environment* 8:614–21.
Stengers, I. 2005. "The Cosmopolitical Proposal." In *Making Things Public: Atmospheres of Democracy*, ed. P. Weibel and B. Latour, 994–1003. Cambridge, MA: MIT Press.
——. 2008. "Experimenting with Refrains: Subjectivity and the Challenge of Escaping Modern Dualism." *Subjectivity* 22:38–59.
——. 2010. *Cosmopolitics I*. Minneapolis: University of Minnesota Press.
——. 2014. "Gaia, the Urgency to Think (and Feel)." *Os mil nomes de Gaia do Antropoceno à Idade da Terra*. https://osmilnomesdegaia.files.wordpress.com/2014/11/isabelle-stengers.pdf.
Stöwe, M., T. Bugnyar, B. Heinrich, and K. Kotrschal. 2006a. "Effects of Group Size on Approach to Novel Objects in Ravens (*Corvus corax*)." *Ethology* 112:1079–88.
Stöwe, M., T. Bugnyar, M. C. Loretto, et al. 2006b. "Novel Object Exploration in Ravens (*Corvus corax*): Effects of social relationships." *Behavioural Processes* 73:68–75.
Steiner, G. 2005. *Anthropocentrism and Its Discontents: The Moral Status of Animals in the History of Western Philosophy*. Pittsburgh, PA: University of Pittsburgh Press.
Strathern, M. 1991. *Partial Connections*. Baltimore, MD: Rowman and Littlefield.
Suárez-Rodríguez, M., I. López-Rull, and C. M. Garcia. 2013. "Incorporation of Cigarette Butts Into Nests Reduces Nest Ectoparasite Load in Urban Birds: New Ingredients for an Old Recipe?" *Biology Letters* 9.

Suchman, L. 2015. “Situational Awareness: Deadly Bioconvergence at the Boundaries of Bodies and Machines.” *MediaTropes eJournal* 5:1–24.
Suddendorf, T., and M. C. Corballis. 1997. “Mental Time Travel and the Evolution of the Human Mind.” *Genetic, Social, and General Psychology Monographs* 123:133–67.
Sussman, A. F., R. R. Ha, and H. E. Henry. 2015. “Attitudes, Knowledge, and Practices Affecting the Critically Endangered Mariana Crow *Corvus kubaryi* and Its Conservation on Rota, Mariana Islands.” *Oryx* 49:1–8.
Swanson, H. A., M. E. Lien, and G. B. Ween, eds. 2018. *Domestication Gone Wild: Politics and Practices of Multispecies Relations*. Durham, NC: Duke University Press.
Swift, K. N., and J. M. Marzluff. 2015. “Wild American Crows Gather Around Their Dead to Learn About Danger.” *Animal Behaviour* 109:187–97.
——. 2018. “Occurrence and Variability of Tactile Interactions Between Wild American Crows and Dead Conspecifics.” *Philosophical Transactions of the Royal Society B: Biological Sciences* 373 (1754).
Szerszynski, B. 2012. “The End of the End of Nature: The Anthropocene and the Fate of the Human.” *Oxford Literary Review* 34:165–84.
Szipl, G., M. Boeckle, C. A. F. Wascher, et al. 2015. “With Whom to Dine? Ravens’ Responses to Food-Associated Calls Depend on Individual Characteristics of the Caller.” *Animal Behaviour* 99:33–42.
TallBear, K. 2013. *Native American DNA: Tribal Belonging and the False Promise of Genetic Science*. Minneapolis: University of Minnesota Press.
——. 2017. “Beyond the Life/Not Life Binary: A Feminist-Indigenous Reading of Cryopreservation, Interspecies Thinking, and the New Materialisms.” In *Cryopolitics: Frozen Life in a Melting World*, ed. E. Kowal and J. Radin. Cambridge, MA: MIT Press.
Tansey, A. G. 1935. “The Use and Abuse of Vegetational Concepts and Terms.” *Ecology* 16:284–307.
Telegraph. 2007. “A Quick Smoke? It’s Good for the Wings.” *Telegraph* (London), June 2.
Telegraph. 2010. “Crows Take a Cigarette Break.” *Telegraph* (London), May 19.
Tengan, T. P. K. 2008. *Native Men Remade: Gender and Nation in Contemporary Hawai‘i*. Durham, NC: Duke University Press.
Terrell, J. 2017. *Offshore: The Sacred Mountain*. Season 2. Honolulu, HI: Civil Beat.
Trigger, D., J. Mulcock, A. Gaynor, and Y. Toussaint. 2008. “Ecological Restoration, Cultural Preferences, and the Negotiation of ‘Nativeness’ in Australia.” *Geoforum* 39:1273–83.
Tsing, A. L. 2005. *Friction: An Ethnography of Global Connection*. Princeton, NJ: Princeton University Press.
——. 2011. “Arts of Inclusion, or, How to Love a Mushroom.” *Australian Humanities Review* 50:5–22.
——. 2015. *The Mushroom at the End of the World: On the Possibility of Life in Capitalist Ruins*. Princeton: Princeton University Press.
Tsing, A. L., N. Bubandt, E. Gan, and H. A. Swanson. 2017. *Arts of Living on a Damaged Planet: Ghosts and Monsters of the Anthropocene*. Minneapolis: University of Minnesota Press.

Tuck, E., and R. A. Gaztambide-Fernández. 2013. "Curriculum, Replacement, and Settler Futurity." *Journal of Curriculum Theorizing* 29 (1): 72.

Underwood, R. A. 2001. "Afterword." In *A Campaign for Political Rights in Guam, 1899–1950*, ed. P. Bordallo-Hoffschneider, 201–13. Saipan: C.N.M.I. Division of Historic Preservation.

U.S. Census Bureau. 2015. *Recent Population Trends for the U.S. Island Areas: 2000 to 2010*. Proceedings from Current Population Reports. Washington, DC.

U.S. Fish and Wildlife Service. 2004. *Endangered and Threatened Wildlife and Plants; Designation of Critical Habitat for the Mariana Fruit Bat and Guam Micronesian Kingfisher on Guam and the Mariana Crow on Guam and in the Commonwealth of the Northern Mariana Islands; Final Rule.*

——. 2005. *Draft Revised Recovery Plan for the Aga or Mariana Crow (*Corvus kubaryi*).* Portland, Oregon.

——. 2008. *Environmental Assessment to Implement a Desert Tortoise Recovery Plan Task: Reduce Common Raven Predation on the Desert Tortoise*. Ventura, California.

Valentine, D. 2012. "Exit Strategy: Profit, Cosmology, and the Future of Humans in Space." *Anthropological Quarterly* 85:1045–68.

van Dooren, T. 2011. "Invasive Species in Penguin Worlds: An Ethical Taxonomy of Killing for Conservation." *Conservation and Society* 9:286–98.

——. 2013. "Mourning Crows: Grief and Extinction in a Shared World." In *Routledge Handbook of Human-Animal Studies*, ed. G. Marvin and S. McHugh. London: Routledge.

——. 2014. *Flight Ways: Life and Loss at the Edge of Extinction*. New York: Columbia University Press.

——. 2015. "A Day with Crows: Rarity, Nativity, and the Violent-Care of Conservation." *Animal Studies Journal* 4:1–28.

——. 2016. "Authentic Crows: Identity, Captivity, and Emergent Forms of Life." *Theory, Culture and Society* 33:29–52.

——. 2017. "Banking the Forest: Loss, Hope, and Care in Hawaiian Conservation." In *Cryopolitics: Frozen Life in a Melting World*, ed. J. Radin and E. Kowal. Cambridge, MA: MIT Press.

van Dooren, T., and V. Despret. 2018. "Evolution: Lessons from Some Cooperative Ravens." In *The Edinburgh Companion to Animal Studies*, ed. L. Turner, R. Broglio, and U. Sellbach. Edinburgh: Edinburgh University Press.

van Dooren, T., and D. B. Rose. 2016. "Lively Ethography: Storying Animist Worlds." *Environmental Humanities* 8:1–17.

Viney, M. 2005. "Eye on Nature." *Irish Times*, May 21.

von Bayern, A. M. P., and N. J. Emery. 2009. "Jackdaws Respond to Human Attentional States and Communicative Cues in Different Contexts." *Current Biology* 19:602–6.

Wadiwel, D. J. 2002. "Cows and Sovereignty: Biopower and Animal Life." *Borderlands* 1.

Warkentin, T. 2010. "Interspecies Etiquette: An Ethics of Paying Attention to Animals." *Ethics and the Environment* 15:101–21.

Wascher, C. A. F., and T. Bugnyar. 2013. "Behavioral Responses to Inequity in Reward Distribution and Working Effort in Crows and Ravens." *PLoS ONE* 8:e56885.

Watson, M. 2017. "Queensland Torresian Crows Form New Nesting Habits in Major Evolutionary Change, Scientists Say." *ABC News Online*. https://www.abc.net.au

/news/2017-01-19/torresian-crows-undergo-major-evolutionary-change-in-queensland/8191076.
Weaver, T. 2010. "Locals Question Military's Plan to Acquire More Land on Guam." *Stars and Stripes*, January 28.
West, P. 2006. *Conservation Is Our Government Now: The Politics of Ecology in Papua New Guinea*. Durham, NC: Duke University Press.
Westphal, M. I., M. Browne, K. MacKinnon, and I. Noble. 2008. "The Link Between International Trade and the Global Distribution of Invasive Alien Species." *Biological Invasions* 10:391–98.
Williams, R. 1980. "Ideas of Nature." In *Problems in Materialism and Culture*, 67–85. London: Verso.
Wilson, C. 2016. "Ornithologist Seeks to Prove Theory NT Desert Hunting Birds Spread Fire to Flush out Prey." *ABC News Online*. https://www.abc.net.au/news/2016-03-03/smart-bushfire-birds/7216934.
Wolch, J. 2002. "Anima Urbis." *Progress in Human Geography* 26:721–42.
Wood, F. 2015. "Tale of a Rascal Crow." *BirdNote*.
Worster, D. 1990. "The Ecology of Order and Chaos." In *Readings in Ecology*, ed. S. I. Dodson, T. F. H. Allen, S. R. Carpenter, et al., 77–89. New York: Oxford University Press.
Wrangham, R. 2010. *Catching Fire: How Cooking Made Us Human*. Perseus Books Group.
Wu, J., and O. L. Loucks. 1995. "From Balance of Nature to Hierarchical Patch Dynamics: A Paradigm Shift in Ecology." *Quarterly Review of Biology* 70:439–66.
Wuerthner, G., E. Crist, and T. Butler. 2014. *Keeping the Wild: Against the Domestication of Earth*. Island Press.
Yancy, G., ed. 2007. *Philosophy in Multiple Voices*. Plymouth: Rowman and Littlefield.
Zalasiewicz, J., M. Williams, R. Fortey, et al. 2011. "Stratigraphy of the Anthropocene." *Philosophical Transactions of the Royal Society A: Mathematical, Physical and Engineering Sciences* 369:1036–55.
Zarones, L., A. Sussman, J. M. Morton, et al. 2015. "Population Status and Nest Success of the Critically Endangered Mariana Crow *Corvus kubaryi* on Rota, Northern Mariana Islands." *Bird Conservation International* 25:220–33.
Zournarzi, M. 2002. *Hope: New Philosophies for Change*. Annandale, NSW: Pluto.

INDEX

CPSIA information can be obtained
at www.ICGtesting.com
Printed in the USA
LVHW042216071221
705551LV00003B/10/J